Stakeholders in Rural Development

Stakeholders in Rural Development

Critical Collaboration in State–NGO Partnerships

John M. Riley

Sage Publications
New Delhi • Thousand Oaks • London

Copyright © John M. Riley, 2002

All rights reserved. No part of this book may be reproduced or utilized in any form or by any means, electronic or mechanical, including photocopying, recording or by any information storage or retrieval system, without permission in writing from the publisher.

First published in 2002 by

Sage Publications India Pvt Ltd
M–32 Market, Greater Kailash, Part 1
New Delhi 110 048

Sage Publications Inc
2455 Teller Road
Thousand Oaks, California 91320

Sage Publications Ltd
6 Bonhill Street
London EC2A 4PU

Published by Tejeshwar Singh for Sage Publications India Pvt Ltd, Laser typeset in 10.5pt. Palatino by Asian Telelinks, New Delhi and printed at Chaman Enterprises, Delhi.

Library of Congress Cataloging-in-Publication Data

Riley, John M., 1953–
 Stakeholders in rural development: critical collaboration in state–NGO partnerships/John M. Riley.
  p. cm.
 Includes bibliographical references and index.
  1. Rural development—India. 2. Rural development projects—Government Policy—India. 3. Non-governmental organizations—India. 4. Non-governmental organizations—Government policy—India. I. Title
 HN690.Z9 C677775 307.1′412′0954—dc21   2002   2001057882

ISBN: 0–7619–9577–3 (US–Hb)   81–7829–083–9 (India–Hb)
    0–7619–9578–1 (US–Pb)   81–7829–084–7 (India–Pb)

Sage Production Team: Aparna Vaidik, Shahnaz Habib, Neeru Nayar and Santosh Rawat

To Ethan and Emma, who bring joy to my life

Contents

List of Tables, Figures and Boxes	8
List of Abbreviations	9
Acknowledgments	11
1. Introduction	13
2. State and Bureaucracy in India's Development	32
3. Non-Governmental Organizations	65
4. Critical Collaboration	87
5. Case Studies	117
6. Collaborative Relationships, Policy-making, and Program Design	153
7. Afterword: The Comparative Context of Collaboration	182
Bibliography	191
Index	209
About the Author	215

List of Tables, Figures and Boxes

A. Tables

Table 5.1	-	NGO-GO Case studies	120
Table 6.1	-	Perceived degree of critical collaboration among the 5 case studies	156
Table 6.2	-	Factors leading to critical collaboration	157

B. Figures

Figure 1.1	-	Factors identified for the study	31
Figure 4.1	-	Levels of NGO-GO interaction	98
Figure 5.1	-	Study Area	152

C. Boxes

Box 6.1	-	Federation of Voluntary Organizations for Rural Development—Karnataka	160
Box 6.2	-	The NPIC in Gujarat	167
Box 6.3	-	MYRADA and the PIDOW Project in the state of Karnataka	175

List of Abbreviations

AWS	Auroville Water Service (NGO)
BINGO	Business-Initiated NGO
CAPART	Council for Advancement of People's Action and Rural Technology (Government of India)
CART	Council for the Advancement of Rural Technology
DANIDA	Department of International Development and Cooperation, Denmark
DHAN	Development of Humane Action (NGO)
DLDB	Dryland Development Board
DNES	Department of Non-Conventional Energy Sources (Government of India)
DRDA	District Rural Development Agency (Government of India)
GEDA	Gujarat Energy Development Agency (state GO)
GO	Government Organization
GoI	Government of India
GoK	Government of Karnataka
GoTN	Government of Tamil Nadu
GRO	Grassroots Organization
GRSO	Grassroots Support Organization
ha	hectare (a measure of land area approximately equal to 2.47 acres)
IAS	Indian Administrative Service (Government of India)
INGO	International NGO
JFPM	Joint Forestry Planning and Management

LPC	Local Policy Culture
MoEF	Ministry of Environment and Forests (Government of India)
MoU	Memorandum of Understanding
NAEB	National Afforestation and Environment Development Board (Government of India)
NCEPC	National Committee on Environmental Planning and Coordination (Government of India)
NGO	Non-Governmental Organization
NPIC	National Program for Improved Chulhas (GoI program)
NWDB	National Wasteland Development Board (Government of India)
PADI	People's Action for Development Initiatives
PHCC	Palni Hills Conservation Council (NGO)
PRA	Participatory Rural Appraisal
PRADAN	Professional Assistance for Development Action (NGO)
SDC	Swiss Development Cooperation (GO)
SHIELD	Sempatti Hills Initiators for Ecological Development (NGO)
SIDA	Swedish International Development Agency
SPEECH	Society for Peoples Education and Economic Change (NGO)
TNFD	Tamil Nadu Forest Department (state GO)
VANI	Voluntary Action Network India (NGO)
WRO	Water Resources Organization (state GO)

Acknowledgments

As in many human endeavors, this book is not the product of a single-handed effort. I have many people to thank for their assistance, support, and contributions, without which it would never have been written.

First, I wish to thank a number of professors throughout my college career who encouraged me to persevere in my studies. In particular, my dissertation advisor, Professor Sue Ellen Charlton at Colorado State University, was invaluable, not only for the tremendous amount of time and effort she put into the editing and review of the (seemingly endless) rewrites and meanderings, but also for her gentle humor and unbounded patience. As an enthusiastic scholar of India, her understanding and support were vital to this effort. My gratitude also goes to professors Valerie Assetto, Shripad Deo, and Jun Xing.

One person without whom this effort would never have been initiated, let alone completed, is my good friend, mentor, and indispensable guide to all things Indian, Sri M.P. Vasimalai. His understanding, kindness and wisdom were absolutely essential to my survival (let alone my studies), from helping me to take my first halting steps in India in 1988, to advising me on the finer points of conducting fieldwork in India, and everything in between. Along with all the good people at PRADAN (now DHAN), he helped kindle my abiding love for India and, especially, her people.

My research was greatly aided by the number of learned scholars at a number of universities, NGOs and research institutes who were willing to discuss my work with me, add their insights, and suggest new ideas and resources that I could use in the creation of this work. These include the following: Subachari Dasgupta of the People's Institute for Development and Training; Arun Sud of CAPART; D.L. Sheth at the Centre for the Study of Developing Societies; Ramachandra Guha and Raghavendra Gadagkar at the Centre for Ecological Sciences of the Indian Institute of Science; Joss Brooks and David Nagle at Auroville; Krishnaswamy Rajivan at the Tamil Nadu Urban Development Fund; and S. Rengasamy of the Madurai Institute of Social Science. Professor D. Rajasekhar, of the Institute for Social and Economic Change, in Bangalore, gave me valuable guidance in understanding and refining the concept of critical collaboration.

I would also like to express my heartfelt thanks to a special friend, Joy Evans, for her support and understanding during the writing of this book.

Finally, it is the Indian people themselves—certainly too many to name individually—who must be thanked. It is through their example that I learned a great deal about being a caring person. They taught me through their every action that there is a wealth in the spirit of humanity that far exceeds the riches of material well-being. I will always be humbled by the warmth that was showered upon me every day in even the smallest kindnesses, which were given without the least expectation of payment in return—which is fortunate, for I could never hope to begin to compensate them for their selfless acts of love.

Chapter 1

Introduction

> In the thirty years that international development has existed as a discrete field of endeavor, only rarely have the diverse sectors within the field been in general agreement on a major topic. This is one of those times, and policy reform or "public dialogue" is the topic. At the moment, a large number of aid agencies all believe that finding ways to alter host country policies is crucial to effective development work in the Third World.
>
> Thomas W. Dichter (1986)

Although the preceding statement was written more than a decade ago, the message it transmits remains an important subject of inquiry in development studies. As humankind witnesses the dawn of a new millennium and the burgeoning of new technologies and scientific advances, and the growth of material well-being for an increasing international middle class, the conditions under which a great number of the world's people live fall below what can reasonably be considered humane. Many of the world's "have-nots" succumb to diseases that have been effectively eradicated or controlled elsewhere around the globe. An alarming number of children die, or suffer mental and physical debilitation, because they do not have access to adequate and sufficient health care, nutrition, or basic education. Families are uprooted from generations-old homes and livelihoods in the name of development and industrialization. Access to productive endeavor, arable land, clean water, health care, shelter and other basic necessities are unavailable to millions.

Such conditions are no surprise to anyone, and a great amount of study, money, and practice has been directed at understanding and addressing their solutions. Massive efforts have been directed and huge outlays of resources have been marshaled by various governments at the national and international levels as well as by educational institutions, private organizations, and businesses in the effort to help the world's underprivileged. Despite these efforts, the conditions for an expanding number of the world's underprivileged are actually becoming worse, not better. Widespread improvements have been made with regard to some problems, while others have received less attention, and the overall level of improvement varies by region and country.

The failure of the massive attempts at development in these countries has led to calls for improvements to, or fundamental changes in, the way that development and underdevelopment have been described and discussed, understood, decided upon and, in the end, actually addressed. There is no shortage of studies, reports, recommendations, and books that bemoan the current situation and suggest that the public and private sectors work in new ways that might improve the work of development. Calls for the various sectors of society to work more closely, in the hope that such concerted efforts may yield greater benefits than isolated activities, abound. Yet, there have been surprisingly few efforts toward inquiry into just how and why such joint efforts actually occur, or how to recognize those attributes and activities of organizations and institutions that make such cooperation likely or possible, such that knowledge of successful efforts can be applied in other places and situations.

This study is a conscious effort to do the latter—to look at actual situations in which a specific kind of relationship have formed. It is a form of relationship that has recently been recognized by some scholars as vitally important to a systemic improvement of the way development is effected, yet not widely practiced or acknowledged. This form of cooperation, a specific type of collaborative effort on the part of governments as well as the voluntary agencies, is not widely practiced in the developing countries (also termed "underdeveloped," "less developed", and jointly as the "Third World"),[1] yet is not unknown there. It involves a conscious effort on the part of both governmental and voluntary organizations to work together, despite deeply engrained mistrust and antagonisms between the two. Such efforts are often volatile, fragile and contentious[2] because such a form of interaction calls for change and cooperation among all involved. This form of inter-relationship is what I am calling *critical collaboration*.

Through a study of what constitutes the important institutional and political factors that allow and/or limit the maintenance of several such relationships in one of the states in southern India, I will attempt to identify those factors that may be applicable in other situations. India contains approximately 40 percent of the world's absolute poor (Duncan 1995: 3) and, as in many developing countries, about 70 percent of its population is rural. India serves as an important place to examine this approach because it has been very active in trying new and innovative approaches to rural development, including forms of inter-institutional collaboration (see Mathiot 1998).

---

[1] A number of titles, including "Third World," have been (and are) used in an attempt to describe a widely heterogeneous amalgam of countries with a variety of social, political, and economic traits that, taken together, place them in a status of "less developed" than the (largely) rich, industrialized western "developed" countries. Despite the biases or inadequacies that accompany the use of any of these terms, it is difficult to proceed without the use of one or several titles; therefore I have chosen to use "developing country(ies)" and the acronym of "less developed country(ies)"—"LDC"—with the recognition (and despite the criticism) that their use may be seen as an unconscious acceptance of the idea of "development" as a discourse (see below).

[2] As John Clark (1991: 50) writes,

> True development is about removing the barriers which restrain people from achieving their full productive capability. This is threatening to those who

Troubled Concepts of Development

Socio-economic development, especially in the world's poorest countries, is an ongoing concern in the West; and there is a widely perceived need to improve the lives and livelihoods of large sections of humanity. To a great degree, it has only been since the close of the Second World War, when many of the developing countries obtained independence, that widespread attention turned to the living conditions of their citizens.

Since the Second World War, a great deal of attention has been given to the idea that these countries could be "developed," a process by which the poorer countries of the world might attain a condition of better livelihood for their people, including improved health care and nutrition, access to education and employment, which in turn would enhance the quality of human lives, and give freedom from oppression and control by others. What it means to be developed or to undergo a process of development (as well as to be less developed or exist in a state of underdevelopment) may not be as easy to define as it first appears. Indeed, there is a growing literature that views the very condition of underdevelopment, and the prescriptions for its eradication, as less a reality of the Third World poor themselves than as a creation of outside observers—a *discourse*,[3] conducted among those in the industrially advanced countries

---

come to see the erosion of their status, and they are likely to fight back. The only socially neutral course of action is to do nothing. Any intervention alters power balances. NGO's business is, inescapably[,] political. It is about taking sides, siding with the poor against those who comprise the barriers to just development.

[3] To understand development as a discourse in the sense (according to Escobar (1995: 39)) that development "created a space in which only certain things could be said and even imagined," it may be illustrative to consider David Howarth's (1995: 119) discussion of the term that "all objects and practices are discursive. In other words, for things and activities to be meaningful, they must be part of particular discourses. This does not mean that everything is discursive or linguistic, but simply that for things to be intelligible they must exist as part of a wider framework of meaning." Thus, those perceptions of what constituted the

who, through their own culturally and professionally conditioned world-views, defined a newly recognized situation (the Third World, underdevelopment) and its solution(s) (modernization, development). As Escobar (1995: 40) points out, modernization—through the investment of capital in the processes of industrialization and urbanization—was viewed as "the only force capable of destroying archaic superstitions and relations, at whatever social, cultural and political cost."[4]

In the earliest post-war years of the development endeavor, both the goals and resultant processes of development were seen as relatively uncomplicated: to create the economic conditions in the poor countries that approximated those of the industrialized West. The assumption was that a rapid and profound change in the economic systems of these countries, an accelerated version of the changes that occurred over several centuries in Europe and the United States during the processes of the Industrial Revolution, would create the foundations necessary for overall social, political, and economic improvements in these societies.

Subsequently, with the failure of the early efforts to produce the degree of expected improvements to the lives and livelihoods of the people of the Third World, there has been a realization that the conditions in these countries are not the result of a single condition, but an array of discrete, yet interlocking conditions (see Chambers 1983). Accompanying the understanding that underdevelopment is the result of multiple, and interrelated causal factors has been the recognition that these factors (and their effects on each other) must be more fully understood and addressed in a more comprehensive manner, if the fundamental reasons for the immiserization of millions of humans are to be countered. It has also been realized that facile, quick and one-step responses have been inadequate. According to a past President of the United Nations University,

---

Third World that found their way into the Western consciousness became the building blocks by which underdevelopment, and the resulting prescriptions for its amelioration—development—were created, defined, and legitimized.
[4] See also Pieterse 1991; Turner and Hulme 1997.

Gone are the early naïve illusions of development as an endeavor in social engineering toward a brave new world. Multiple goals have now replaced the initial single focus. There is now a greater understanding of the profound interaction between international and national factors in the development process and an increasing emphasis on human beings and the human potential as the basis, the means, and the ultimate purpose of the development effort (Soedjatmoko, quoted in Todaro 1985: 61).

There has been a hot debate among both scholars and practitioners for decades over issues concerning how the process of development in the poor countries should take place, which includes the normative questions as well as the practical considerations regarding how development as a *process*, or means, can lead to a development as an *end*, or result. However, the measurements of levels of poverty, illiteracy, deaths due to easily curable illnesses, under- and unemployment, as well as other indicators of underdevelopment continue to largely belie the predictions of progress. In fact, the very notion of what development *is*—what it means for a country, or a village, or an individual to be developed,[5] and what comprises the minimum as well as the optimal requirements for development to have occurred—has been in question throughout the past several decades, and is still subject to differences of opinion.

Natural Resources and Rural Development

The issue of the effect of natural resource use (and misuse) on human well-being, as well as the long-term protection of the natural environment, is inextricably linked to the economic and social development in the Third World. The lives and

[5] Lacey (1991: 213) discusses the difficulties of defining development: "'Development' may denote the moral aspirations of an impoverished people; it may also represent (depending on the perspective from which it is used) the actual achievement of certain societies, the alleged necessary trajectory of all viable societies, or what the agents of powerful economic institutions regard as the only realistic possibilities."

livelihoods of citizens, the majority of whom live in the rural areas, are often directly dependent on their natural surroundings. Not only is arable land necessary, but the countryside is often a major and immediate source of sustenance, building materials, medicines, and other products essential to the rural poor. However, at the same time, those same resources are valuable for private gain and for wider development purposes as well, especially in countries that often have few other marketable commodities. As a result, the question of who is allowed access to primary resources becomes a vital one.

Contending pressures for the use of existing resources create a dilemma regarding their management. Historically, governments have assumed responsibility for resource control and management, frequently preventing access by rural populations to much needed resources under the guise of controlling their use for the purpose of national development (see Ascher and Healy 1990; Gadgil and Guha 1995). In the name of sustainable development and sustainable rural livelihoods, governments (at several levels) are today increasingly attempting to make these resources more available to the rural poor while, at the same time, allowing their use by commercial concerns for larger development aims. All the while, they are hesitant to relinquish any of their long-held control of the same. The question of equitable use of natural resources, therefore, is central to both rural development and to the long-term sustainability of the natural environment.

Governments cannot respond to every local circumstance of contention over resource use even if such a balancing act were possible. National policies, in particular, are often insensitive to the often minute, yet crucial, variations in each rural locale to be effectively and equitably applied. This is because the governments simply do not have the personnel or other assets to respond to each individual and unique case. It is for this last reason, among others, that indigenous nongovernmental organizations (NGOs) have been heralded as an alternative solution to the problems of inequitable resource use. While most indigenous NGOs lack the resources, size, authority or expertise to solve widespread problems on their own, they are seen to have attributes, which are otherwise lacking in governments, that are important to rural development. In combination with their governments' greater resources and scope, it is suggested, the successes of numerous small NGOs can be replicated and multiplied to bring about large-scale change.

Recent literature on development suggests that it is important that NGOs take an active role in influencing policy officials in regard to governmental attitudes and practices toward development and regarding the poor citizens of their countries. There is a widespread conviction that NGOs are, in some ways, better qualified than government organizations (GOs) to successfully implement development programs and projects, and that their inclusion in the policy process can be an important addition to governments' development activities.[6] Despite these exhortations, there exist only a few, practical studies discussing how such participation actually operates.

In his study of NGO-GO interrelationships, John Clark (1992) identifies several approaches that a NGO may take in working with a government body, ranging from straightforward cooperation with government agencies to active efforts at thwarting the state's activities. According to Clark, an NGO can choose: 1) to forego any sort of involvement in the government's program; 2) to complement the state (by developing its own program to enhance the government's existing services); or 3) to attempt to reform the state through deliberate collaboration with the pertinent government agency to improve the formulation and delivery of services. Of these, the latter approaches—complementing the state and reform—appear to be constructive forms of relationships and may be the most likely to foster an atmosphere in which both policy-making and implementation can be made more effective. The third, collaborative approach, is the focus of this study.

---

[6] Clark (1991: 107–8) writes that, "The NGO experience is almost wasted unless it is used to change the attitudes and practices of decision makers."

The Study: Critical Collaboration

Despite numerous studies that indicate the superior performance of some actors outside the official development sector and which call for an increased role for those actors in the creation, planning, and implementation of rural interventions, development continues, in many cases, to be closely controlled by governments. How is it then possible to expand the role played by actors outside the government sector in the making and execution of development interventions?

The book examines the nature of a specific form of relationship that takes place between non-governmental organizations (NGOs) involved in environmental and natural resource issues, and government organizations (GOs) in the state of Tamil Nadu, India—a relationship in which both act together in the design and implementation of rural development programs. This form of interaction is more than an agreement between two organizations to cooperate or complement each other in the execution of an existing program. Rather, it is a form of cooperation between an NGO and government agency that involves the acknowledged, active participation of both parties with each other in one or more aspects of a particular issue. This form of collaboration is termed critical collaboration because the NGO is specifically empowered to more fully participate in the process, and acts in association with a GO and often as a *critic* of the GO.[7]

Such relationships allow both government agencies and NGOs to work together as partners, or "stakeholders"—parties who have an interest in the outcome of a particular problem, and who are directly influenced by the actions taken by each other to solve the problem (see Gray 1989)—in both making and implementing efforts at rural development. The nature of this form of relationship is one in which an NGO has an opportunity to participate in the decisions that shape the way the

[7] In their foreword, Reddy and Rajasekhar (1996) cite M.K. Bhat, who, to the best of my knowledge, coins the term: "While the government expects NGO involvement within the socio-economic matrix already laid down, NGOs would like to have a *critical collaboration* with the government" (emphasis added).

development interventions are conceived and planned, as well as the way in which they are conducted at the implementation stage (NGOs are often sought to carry out pre-designed programs due to their supposed advantage in understanding individual local conditions, yet are not as often consulted in the overall formulation of policies or the design of programs; see Chapter 4). Such relationships, although not widespread, do occur, not only in India but in various other developing countries as well.

The objectives of this study are to identify some of the key factors leading to the establishment of relationships characterized as critical collaborations between NGOs and government organizations; to examine what forms those efforts take; and, through an understanding of the features of the working relationships, to identify those characteristics of the individual NGOs that are essential to the creation of such relationships.

The Concept

Although recent NGO-government relationships have been characterized by mutual mistrust and hostility, NGOs and GOs do work with each other under certain circumstances. Each are perceived to have certain intrinsic characteristics that are deemed necessary, if not sufficient, for the creation and successful implementation of development programs and projects, and for the replication of such successes in multiple locations and differing circumstances. A number of authors posit that the deliberate association of GOs and NGOs in collaborative interrelationships in ways that allow them, together, to use their own best resources and capacities to mutual advantage, will result in a number of improvements to the current status of development efforts. This is the essence of what I have termed critical collaboration, following Reddy and Rajasekhar's (1996: 10) assertion that

> linkages with the government should not be uncritical.... NGOs have to play a dual role—they need to collaborate

with the government, and at the same time, they have to oppose those government policies which are detrimental to the poor.... [T]he relationship with the government has to go beyond linkage or interaction and should function on the basis of working together in a *mutually interdependent fashion* (emphasis added).

The concept of critical collaboration is at the heart of this investigation—a working relationship that allows each involved party the use of its distinctive characteristics in fashioning and implementing rural development interventions while, at the same time and in the same program/project, sharing the resources, abilities, and decision-making responsibilities of its counterpart/s. The point of such a relationship is that the NGO—as well as its GO counterpart—has the ability to view the policy or program critically and to make its concerns, experiences and opinions a real component of the policy decision-making process.

One objective of the inquiry is to gain an understanding of what comprises critical collaboration, through an examination of the components of several cases. Thus, both the salient factors that make possible the establishment of such a relationship and views of the participants in that relationship will be studied.

The Central Question

In understanding the nature of critical collaboration between NGOs and GOs, it is important to look at several questions. The central question is: under what conditions—the identifiable political, organizational and social contexts in a given situation—do government organizations (GOs) and non-governmental organizations (NGOs) work together as interactive and interdependent entities in together formulating and delivering rural development interventions?

The emphasis on understanding a specific form of relationship is deliberate; if, as many argue, non-governmental organizations *as such* have distinct attributes that actually make them superior in some ways to governmental agencies in bringing about real improvements in peoples' lives, one may suggest that only when an NGO has the freedom to use those attributes in a meaningful fashion within the context of a working relationship with government agencies can such a relationship truly exist.

The Supporting Questions

In order to answer the central question, important characteristics of both the actors and the surrounding circumstances must be identified. Questions aimed at determining such factors include:

- What characteristics of NGOs and government agencies are important to the process of creating relationships of critical collaboration? Are these characteristics inherent to the NGO and government sectors as a whole, or do they exist only on an individual, idiographic basis? To the extent that similar characteristics are inherent in a wide range of NGOs and GOs, it may be possible to understand the possibilities of such relationships existing elsewhere.
- What attributes of the social, political, and economic environments in which such relationships take place are influential in the creation of a relationship of critical collaboration? Can the existence of such a relationship be identified?
- What actions did the parties deliberately take that allowed, enhanced or impeded the establishment of a critical collaboration?
- In any individual case, do there exist unique intervening factors that may be decisive in making such a relationship possible?

Indicators of Critical Collaboration

In finding answers to the questions posed, a number of important characteristics of NGOs, GOs and the contexts within which they operate must be identified and investigated. These include the following:

Influencing Factors

- The attributes of NGOs. Perceptions of the characteristics of a particular NGO which seeks a collaborative relationship with a local or state-level GO in a rural development effort are central; the present and past activities of an NGO, as well as its area(s) of expertise, will affect the way it is perceived by a GO as a possible partner for collaboration. Also, the stance that an NGO takes toward local governments, and what they view as their role in development in general—referred to as their ideology— may affect government decisions to work together.
- The attributes of GOs. The characteristics of a government agency will also affect both its willingness to work with, and its acceptance by an NGO as a working partner. Some agencies are known to be hostile toward working with outside individuals and organizations, while others have a history of cooperating with NGOs. Does the agency have the jurisdiction to accomplish what the NGO desires or the discretion to make changes in established policies? Is there a particular individual interested in a particular issue or in collaborating with NGOs?
- The contexts shaping the relationship. The political, economic and social environments in which rural development efforts ultimately occur are mediating factors; the possibilities for successful interrelationships are affected by the nature of these environments. For instance, the degree to which particular local government agencies or NGOs are accepted by the rural populations

and are able to work with local elites and village groups are all pivotal to successful efforts. In addition, the presence and influences of foreign donors and organizations often shape the direction efforts take. An investigation of these factors should make it possible to identify the resulting interrelationships between NGOs and GOs.

Outcomes

- The nature of the relationships. The degree to which each relationship can be characterized as a collaborative one is investigated, especially the degree to which the NGO acted as a critical partner in the collaboration.
- How the relationship was established. The particular events and circumstances that allowed, or fostered, the creation of a collaboration between NGO and GO are presented. Do all collaborative relationships result from a common chain of events, or do distinct paths of relationships lead to similar results?

The initial question of this study also involves how and why the critical collaboration has occurred: the reasoning for the decision to collaborate (and the circumstances leading to the decision to collaborate); the circumstances under which contact between the NGO and the GO was initiated (initially cooperative or antagonistic); whether the collaboration was initiated by the government, the NGO, or by a third party; and the relevant social and political contexts in which the relationship was initiated. An understanding of the factors underlying the decision to collaborate is important, in that at some point all concerned parties made a conscious decision to enter into what is a potentially difficult working alliance. In critical collaboration, both parties must agree to acknowledge and consider the concerns and ideas of the other/s.

Methodology

Approach

The fieldwork for this study was conducted in the south Indian state of Tamil Nadu during 1996 and 1997, and is based upon a number of case studies involving examples of collaboration between NGOs and GOs in that state.

I identified a number of NGOs in several regions of the state of Tamil Nadu and conducted preliminary interviews to identify organizations which have had the opportunity to work with one or more corresponding state or local GOs in programs or projects involving environmental and natural resource issues. There was a wide range of responses to the initial inquiries: several NGOs had not worked with GOs in any cooperative manner (let alone collaboratively), or did not wish to do so; a few were clearly less interested in the development of the rural poor than in their own; others did not wish to participate in the study; and a few that had worked with state agencies had done so as contractors in a pre-arranged program. The five NGOs chosen for study are each distinct in several ways: although the NGOs differ in a number of characteristics and behaviors, each case was representative of what may be seen as either an existing or a nascent form of critical collaboration (which will be discussed further).

The study was restricted to the state of Tamil Nadu. Individual state governments in India vary both in their stance vis-à-vis the central government and national policies, and toward NGOs operating within their borders (see Chapter 2). The conscious choice of limiting the study to a single state was made in order to reduce the possibility that state differences would unduly distort the variables under study. Interviews were conducted in the English language with NGO and GO officials, as well as a number of Indian scholars at various universities, institutes, and research organizations. No interviews were conducted at the village level, although a number of extended visits to villages with both GO and NGO officials, and discussions with villagers, enhanced my understanding of the cases and aided in gathering information. As a result, the findings of this study should be interpreted as indicative of only a small portion of the NGO sector involved in India's rural development process.

Gray and Wood (1991) identify three components of collaborative associations: 1) the preconditions which make the creation of collaborations possible, and which motivate groups and individuals to participate; 2) the process of the collaboration itself; and 3) the outcomes of the collaboration. This study focuses solely on the first. Both processes and outcomes are complex and often controversial in terms of their accuracy and relevance, and may require years of measurement and analysis. Without answering the questions of whether the NGOs and GOs working in concert have sustained their collaborations, or achieved successful conclusions, this book looks at the dimensions of NGOs, GOs, the political, social and other contexts within which they operate, in the attempt to understand the creation of actual collaborative alliances.

Organization

The book is organized into six chapters. Chapter 2 views the nature of the Government of India (GoI) and its subsidiary state and local counterparts in the context of the country's environmental and natural resource problems. Constitutional, political, legal, and bureaucratic components of the evolution of environmental concerns are briefly discussed to elucidate the political background of the study.

Concern for the natural environment, and the use of natural resources in India is not merely a recent issue; it extends, in modern history, to the early years of the British presence on the subcontinent. The chapter will include a brief review of the history of environmentalism in India to build an understanding of the historical antecedents that have led to the present status of the environment and Indian responses to that situation.

Chapter 3 assesses the other set of actors involved in the study. It consists of a literature review of non-governmental organizations (NGOs)—how they are perceived, their history, including a discussion of their past and current role in development, and some of the characteristics and activities of NGOs that color their relationships with governments—including the possibility of forming collaborative alliances.

With a basic understanding of some relevant attributes of the two sets of actors, Chapter 4 examines the nature of NGO-state interactions, in the past as well as in the present. In particular, the evolution of the interrelations in the Indian context is reviewed.

Private charities, religious missions and other voluntary organizations have been active in India for centuries. Until recently, their activities were considered ancillary and basically unrelated to the task of large-scale development, long assumed to be the purview of the government. In the past several decades, however, NGOs have widened the scope of their activities; many are politically active today, and have assumed a number of functions formerly monopolized by their governments. The purported attributes of NGOs are manifold; ultimately, they are seen as an effective alternative to government-led development interventions. Perceptions of the attributes and abilities of NGOs are presented and examined. In addition, the nature of Indian NGOs, and the political and social contexts in which they operate, are explored.

The chapter includes a review of what are seen as possible working interrelationships between GOs and NGOs, including government expectations of the ways in which their indigenous NGOs can work with them, as well as NGOs' responses to such expectations and attitudes.

In Chapter 5, individual case studies of NGO-GO interactions are presented in narrative form. The studies focus on the nature of the NGOs and their GO counterparts (at several levels, from local to national), the evolution of the particular cooperative relationship that is under study, and the nature of the relationships at present. Individual and important characteristics and activities of the organizations are highlighted, and their progression toward possible collaborative relationships will be explained.

Chapter 6 is an analysis of the factors that have been identified as important to the process of creating collaborative relationships between the actors in each of the cases. This chapter will present the findings of what occurred in the establishment of the relationships between the NGOs and the government functionaries. The cases are examined in detail, according to the factors that have been outlined in the first chapter; they will be compared and contrasted with each other. In particular, the degree to which each NGO was successful in entering and maintaining a collaborative interrelationship with its GO partner will be assessed.

The Afterword offers suggestions for the creation of NGO-GO critically collaborative relationships. If collaboration is to become more than occasional, isolated occurrences—to become a truly significant force in rural development—those attributes of both the voluntary and government sectors necessary to allow, or facilitate, interdependent collaborative relationships to occur and expand must be identified and understood. What are seen as several of these attributes of both NGOs and GOs will be discussed.

The purpose of this undertaking is straightforward. It is an attempt to provide other scholars and practitioners of rural development with some evidence that there exist today new and largely (although not completely) untried methods of improving the conditions of the huge number of rural poor in the Third World for whom the social, political and technological advances made elsewhere are yet unavailable. It is an endeavor to provide some concrete responses to those who identify some of the barriers to effective rural development in LDCs, as well as their causes, yet too often do not look beyond the realization to seek solutions to overcome those barriers. Finally, it is an attempt to identify some of the attributes necessary for individual organizations to enter into such relationships.

Figure 1.1
Factors Identified for the Study

Influencing Factors	Indicators
Non-Governmental Organizations (NGOs)	size; age; experience (including relationships with other GOs and NGOs); expertise; approach to working with GOs
Government Organizations (GOs)	Willingness to work collaboratively with NGOs; jurisdiction over resource in question; discretion to change policies; individual involvement
Contexts shaping the relationship	National, state and local governmental policies; outside influences (including elite, funding agencies); history of NGO-GO relations; local policy culture (LPC)

Outcomes	Indicators
Nature of the NGO-GO relationship	The existence and extent of a working relationship: the degree to which the relationship is actually perceived by the parties as a critically collaborative one
Approach to the relationship	Whether the particular approach to working together has led to a critically collaborative relationship

Chapter 2

State and Bureaucracy in India's Development

> Third World governments are seldom in a position and often are not disposed to assist the rural poor, who are mostly unorganized, "out of sight and out of mind," having little political clout. For the most part, the poor must cope on their own. Governments, donors, and intellectuals who espouse the current orthodox, neoliberal model of development are not unconcerned with poverty, including rural poverty. ...[T]here is a lingering hope, not supported by evidence, that economic growth will gradually eliminate poverty and put the national economy on a firmer foundation for long-term and equitable productivity.
>
> Norman Uphoff et al. (1997)

As concerns over development and underdevelopment surfaced with the post-war emergence of newly independent states, a wide range of economic and other problems in former European colonies of Asia, Africa and Latin America gained widespread attention. These problems formed the basis of a way of looking at social, political and economic conditions in a widely diverse number of countries and, as a result, triggered efforts on the part of Western countries for their eradication. Questions arose over such issues as the most efficacious and (politically and socially) acceptable means of achieving development ends, the sector of society best equipped to effect improvements for the largest number of the poor over the long term, and how such improvements can be accomplished in the most equitable manner. Over time, these perceptions shifted, especially as favored approaches failed to meet the expectations of donors, policy-makers, and planners.

The Government of India, like those of many other developing countries has, over the past several decades, undergone substantial changes in its approach to development. The country's initial approach, which placed the majority of the control of planning and implementation in the hands of the central government, has been amended to include other actors in the process. The degree to which the government's changing development policies have been successful in decentralizing the development process, however, has not been uniform or widespread. At least in part, the degree to which the conditions of the rural areas and their residents have improved has varied with individual states' interests in bringing about such change (see Drèze and Sen 1996). Entrenched interests, as well as existing institutions and longstanding practices, have also been highly resistant to the changes. While there are additional obstacles to the successful accomplishment of development efforts in India, this chapter focuses on some of the constitutional, bureaucratic and institutional impediments that shape and limit the progress of rural development in India.

This chapter begins with a general review of the state as the force behind national development and particularly, with regard to environmental issues as they relate to the development process. This is followed by a discussion of India's environmental concern and responses. The chapter concludes with a look at some of the constitutional provisions, and institutional and bureaucratic features that affect the development process in India.

Governments and Development

One widely held critique of many developing countries concerns the degree to which many governments view themselves as the sole and rightful leader of their country's efforts at social and economic development. Many developing countries gained independence within the past five decades, and have had limited experience with democracy; governments often see their presence and their (sometimes autarchic) actions as necessary to overcome the parochialism and atomization of different cultural and ethnic groups within their borders, in an effort toward nation-building and development (see Esman 1991: 91). Although there has been a growing movement towards more democratic forms and practices, many governments still retain, to varying degrees and in some areas of governance, some of their historically authoritarian elements. While it is certainly an oversimplification to attribute undemocratic policies and activities to LDCs as a whole, the continuing presence of autocratic policies and practices (which may be vestiges of colonial rule) are common to a number of countries.

The failure of many governments to place a high priority on the welfare of their impoverished citizens is evident. Often their overriding concern is simply their own ongoing survival. Other, less immediate goals are subject to this imperative. As a result, development policies and objectives are shaped such that, even if they do not serve to enhance the control of the ruling regime, they do not compromise it. The tendency of a number of national leaders is to centralize power in their own hands, instituting a wide range of rigid administrative procedures, prohibitions and regulations, and otherwise limiting the powers, responsibilities, and autonomy of subsidiary governments and other groups in the name of order and stability.

In their drive to consolidate and centralize policy-making and operational control in the name of the state's overall welfare, governing elites often threaten the very legitimacy they seek through the imposition of controls. Most LDC governments lack sufficient resources to follow through on the powers they claim and cannot hope to expand the scope of existing efforts to their entire populations. In addition, even the most sincere efforts are often limited in their effectiveness, given the social, cultural, political, geographic and other variations within national borders, as well as the enormity of the task. As a result, the inability of governments to fulfil their promises undermines their peoples' support and weakens their legitimacy.

The primary role of the GoI in leading and guiding the direction of development has been a major feature of its policies since India's Independence in 1947, and continues despite numerous efforts at decentralization and privatization of significant areas of the country's economic sector. At the time of independence, the debate over the direction of Indian society and the economy centered on three visions of the country's future: a Gandhian view of a return to a largely rural, self-sufficient and decentralized polity and economic system based on India's villages; a socialist ideal, promoted by Nehru, in which the government would take a leading hand in planning and controlling the creation of a centralized, industrial state that would bring India into the modern world; and a liberal, capitalist vision favored by Sardar Vallabhbai Patel. The policy that was eventually adopted by the GoI was based largely on Nehru's vision (although concessions were made to the proponents of the other schools of thought in the name of national consensus-building) and focused mainly on the modernization of India through industrial growth and urbanization. Until the years following Nehru's death in 1964, relatively less attention was given to the rural poor who, altogether, account for over 70 percent of the labor force and 80 percent of India's population (Rudolph and Rudolph 1987: 312).

Rural development policy (which still focused on agricultural production) took a sharp turn with the introduction of the technologies of the Green Revolution to Indian farmers. The advances were widely adopted in some areas of India, especially among larger land-holders and others capable of investing in the inputs necessary to successfully cultivate the high-yielding varieties of wheat and rice which were at the basis of the "Revolution." However, the required inputs (seeds, chemical fertilizers, and dependable supplies of irrigation) were inaccessible to the majority of India's marginal and landless farmers. While large-scale grain production increased, the Green Revolution also resulted in a widening of the gap between rich and poor in rural India, as in other countries (see Ascher and Healy 1990).

Uneven increases in agricultural and rural economic growth that were exacerbated by the limited access of poor farmers to the technological innovations of the Green Revolution led in the late 1960s to a shift in the focus of rural development efforts. With the realization that both the industrialization strategy (which had led to significant, although imbalanced, urban economic expansion) and the experience of the Green Revolution did little to improve the lives of the majority of India's rural poor, Indira Gandhi (daughter of Jawaharlal Nehru and Prime Minister of India, 1966–77 and 1980–84) attempted to address the problems of uneven development and economic power by creating programs of income redistribution and accelerated growth, designed to assist specific target groups among the rural poor (Rudolph and Rudolph 1987: 326).[1] The focus on poverty reduction as the principal impetus for rural improvement, primarily through job creation and increased investment in agriculture, has been the mainstay of India's rural development policy to the present.

Terry Alliband (1983) posits that India, as many other countries, has followed two fundamentally different approaches in the way that rural development is carried out: "rural development" versus "community development" (see also Ali 1992). Community development, a widely popular concept during the 1950s and early 1960s,

> aims at improving the overall community infrastructure, with particular emphasis placed on a community's *human* resources. The basic aim of the community development approach is to generate community-based, community-wide, problem-solving competency (Alliband 1983: 5). [emphasis as in the original]

[1] Hardgrave and Kochanek (1993: 379) write that "In emphasizing growth per se rather than as a development plan of 'growth with justice', India opted for production without social change, a policy that implicitly accepted the growing gap between the 'haves' and the 'have nots'. As contradictions within the economy deepened, Prime Minister Indira Gandhi gave increasing rhetorical emphasis to the goals of social justice."

In contrast, rural development, which has dominated more recent development policies, places its first priority on economic growth. To foster this, rural development programs focus on the establishment and enhancement of infrastructures that facilitate economic growth under the assumption "that rural entrepreneurs will take advantage of economic opportunities if suitable wealth-generating facilities are available" (Alliband 1983: 1). The assumption is that improvements in local economies will translate into increased human well-being.

According to Alliband, the shift from community development to rural development took place for several reasons: a lack of political commitment to fundamental rural change; the diversion of necessary resources for urban needs (abetted by the disproportionate power and visibility of urban versus rural interests); and a scarcity of allies for the cause of the rural poor. This was enhanced by the relative ease of allocating resources to "progressive farmers" eager to accept modern government approaches and programs, rather than undertaking the more difficult and time-consuming processes of interacting with, and increasing the "competencies" of entire communities to accept and adjust to major changes (Alliband 1983: 7).

Autar Dhesi writes that India's development policy today reflects a return to the focus on community action and participation. In its Eighth Five-Year Plan (1990–95), the GoI recognized the importance of "'human development' as a major objective of development. The plan explicitly recognized the importance of decentralized broad-based development, and people's initiative and the role of NGOs in development is given due recognition" (Dhesi 1996: 212). It also states that despite the professed concern and the GoI's interest in including outside actors, including NGOs, in its official rural development policy process, few actual changes have been made in the development process.

The extent to which policies formulated at the centers of national governments are actually implemented at the local level depends, to a degree, on the operational framework of those governments. In small countries with unitary forms of government, policy- and decision-making are often the responsibility of a small, concentrated number of individuals and groups; the execution of such decisions is carried out by subordinate bodies who are intended to execute the will of the center. However, in countries characterized by large and dissimilar populations with greater social and geographic complexity, the administration of central decisions becomes unwieldy. The creation of multi-level governments, with both shared and separate powers, is designed to address some of the less tractable differences among large and diverse populations, especially those issues that may lead to inequitable law-making and governance under unitary control. At the same time the functions of such governments, in which the constituent parts actually function (as opposed to local bodies which are created solely as "window-dressing"), become considerably more complex and difficult. In such situations for the purposes of fair and equitable governance, multiple layers of government are often created.

Federalism

India, unlike many other developing countries, has an immense population, representing a multitude of distinct cultures, religions and languages. The difficulties of governing such a large and heterogeneous citizenry are great, especially in the context of creating a democracy where none had existed previously. To do this, the Founders of the Indian Constitution adopted a federal form of government. The primary rationale was to create and maintain, while allowing for the expression of this profusion of diversities, the idea of a unitary national identity and a semblance of a continuing, coordinated, functioning democratic political entity.

Federalism was seen as a practical solution to two problems at the time of Independence. The first, a direct result of British colonial rule, was that of integrating with the British provinces those portions of the subcontinent that had not been subject to colonial rule: the princely states. These states (562 of which existed at the time of Independence) were largely autonomous;

their integration into the Republic was seen by the Founders as necessary to statehood. Another, less tractable, difficulty stemmed from the presence of a significant minority population of Muslims, which feared subjugation to a Hindu majority. The formation of a multi-level government with appropriate safeguards for multiple, social, ethnic and religious variations was chosen by the Founders.[2] However, the form of federalism that was created by the Founders of the Constitution decidedly favored the center; rather than giving the federal and subordinate governments exclusive powers, the Central government has a disproportionate share of political power and dominance over individual states.

Before 1967, center-state relations in India were relatively unproblematic. The charismatic leadership of India's first Prime Minister, Jawaharlal Nehru, and the dominance of the Congress Party in almost all of the states (thus facilitating cooperation between and among both center and state governments) fostered relatively amicable relations. The death of Nehru in 1964, and the breakdown of the Congress hegemony in the elections of 1967, however, led to more complex and less harmonious center-state relations. Caste-based and regional political parties proliferated, and replaced Congress rule in a number of states. At the same time, mass franchise based on regional and cultural group aspirations began to flourish, extending participation in political processes from the upper and middle levels of society (who had been socialized into national politics with the coming of Independence) to the urban and rural poor. The conflicts that previously had been quietly settled among Congress party leaders subsequently became the subject of

---

[2] In a pre-Independence attempt to address these problems the British introduced, in the Government of India Act of 1935, a set of political reforms leading to decentralized government. More symbolic than realistic in terms of practical application (many of its provisions were never implemented), the Act was nevertheless the first move away from the British tradition of centralized control and toward a federal system for India. Despite its limitations, according to Hardgrave and Kochanek (1993: 127), the Act "established the principle of federalism for the Indian subcontinent.... The idea of federalism seemed to be taken for granted by those who drafted a constitution for a free India."

public negotiations between the Central government and the states (see Nirmal Bose 1987: 36). The differences among the individual states of India and the proliferation of regionally-based political parties and state governments has led to calls for greater state autonomy, including the decentralization of planning, greater independence for state control of those subjects on the Constitution's State List, and changes to the apportionment of national resources. There are significant differences among the states with regard to their levels of development and growth, and individual states are engaged in persistent struggles with both the Central government and each other to redress those disparities in their favor.

The result of these developments is, according to a number of authors, arguably quite similar to the intent of the United States' Founding Fathers—the accommodation of two seemingly contradictory forces: the desire of the federating units for national unity and, at the same time, retaining their individual independence. While political power in federal India is constitutionally weighted toward the Central government, the nature of the relationships between the levels of government have created, according to some, the sense of a functional federalism. As Verney and Frankel (1986: 572) see it,

> [a]lthough India was not established as a federal system on the North American model, enormous changes have taken place in Centre-state relations since 1947. So great has been the transformation that the federal character of the constitution...is now taken for granted.

Referring specifically to the area of development, Paul Appleby (1953: 54) writes that India is "most federal" because of the Central government's reliance on the individual states for implementation of national development and plan policies. This sentiment is not shared by all. For instance, S.R. Sen (1988) argues that since Independence, the trend toward centralization of political control in India has far exceeded the intention of the framers of the Constitution. Tummala (1994: 71) posits that although the original intent was to create a federal form of government, certain constitutional provisions allow the expropriation of state power by the Central government, in effect reducing the government "virtually to an unitary form." As Nirmal Bose elaborates,

> The problem of the Union-State relations is really very serious. The states have practically no existence of their own. For almost everything—political, economic, cultural—they are entirely dependent on the Centre. The states have been brought down to a mendicant status.

The status of Indian federalism remains controversial. There is a widespread feeling that federalism in India has become so unevenly dominated by the center that some remedy must be taken if India is to continue as a federal union. However, there is no consensus regarding the sources of the problem or its solutions. A number of scholars see the problem as a financial one, and call for a redistribution of tax revenues to enable greater state autonomy (Ray 1987). Mukhopadhyay (1987) points to the rigidity of the central bureaucracy (especially the All-India services) and its domination of state-level administrations. Especially in light of the Central government's focus on increasing grassroots democracy and *panchayati raj* (literally meaning 'rule of five'), he favors greater autonomy for state-level governments. Rasheeduddin Khan (1997: 16) argues that India's unique federal polity is based on "the existence of distinct and viable cultural-cum–socio-economic regions, competing for priority of allegiance with the in-built and felt consciousness of a unified Indian nationality." In his view, the expression of these differences is important to the creation of a national identity; he suggests a re-organization of states into smaller, more homogeneous and politically manageable units that would lessen the inequalities that exist between unequal groups and minorities within states, make government more responsible to local demands, and increase participation in development.

Given the continuing arguments over the nature of Indian federalism, and the institutions and processes created in order to foster a sharing of powers in India, it is instructive to examine what they are and how they operate, especially as they apply to rural development. The composition of the Constitution of India—the division of powers between the central and state governments, and the way those powers have been used by the different levels of government—affects the way that development is administered, and thereby the dynamics of inter-organizational relationships.

Constitutional Provisions

In comparison with other federal arrangements, the Indian Constitution contains several striking features. An important distinction is that the Center and states operate under a single Constitution. This national Constitution delimits the powers of both states and Central governments in the Seventh Schedule, in which three "lists" specify the activities of national and state governments under normal conditions. The Union List originally designated ninety-seven areas of authority exclusive to the Central government, including defense, foreign affairs, banking, and income taxation. The sixty-six items of the State List are, under normal conditions, the sole authority of the individual states, while the Concurrent List contains forty-seven areas over which both the Center and the states share jurisdiction. The majority of government responsibilities specifically involved with rural development, natural resources, and the natural environment, such as agriculture (including education, research, and taxation), fisheries, and water resources, are constitutionally reserved to the states.[3] Neither the Union List nor

---

[3] According to Sunil Khilnani (1997: 36), the decision to place matters of rural development under state jurisdiction was a political concession on the part of Congress toward the powerful rural landowning constituency upon which it depended. Given the importance of the rural elites, "[the Constitution] left the matter of social and economic reform...to the regional legislatures, where the landed elites could bring local pressure against redistributive moves" (For a discussion of the role that rural landholders played in India's independence movement, and their continued importance to Congress after Independence, see Manor 1990).

the Concurrent List, as originally enumerated, contained any powers specifically involving rural development, although this has changed.

Despite this segmentation of responsibilities, the Constitution contains several major features that give the Central government considerable power to override state decision-making. First, if conflict occurs between a state and the Center regarding an issue on the Concurrent List, the latter prevails. The Emergency Powers (granted under Articles 352, 356, and 360) enable the Central government to transform India into a unitary state under conditions of threats (respectively) to national security, the failure of the constitutional machinery to operate as intended in a state, or a threat to financial security. Under Article 249, the national Parliament is granted special legislative powers over any item in the State List for a specific period of time if relevant legislation is called for on a national scale. Finally, under Articles 256, 257, and 365 the Center can assume powers to direct state governments to comply with Union laws and, failing compliance, levy substantial penalties. If individual states continue to ignore the Center's directions the latter may, under Article 356, take over the governing of the former.[4] The unequal division of power even gives the Center the legislative power to create new states, alter the boundaries of existing ones, and even to abolish a state. Parliament may by a simple act of legislation "admit into the Union, or establish, new States on such terms and conditions as it thinks fit" (Government of India 1996: Article 2, p. 2).[5]

---

[4] Balveer Arora (1987: 176) writes that the Central government has used this constitutional remedy—called "President's Rule"—in numerous cases to replace state governments which "[have] chosen to pursue independent policies" with more "compatible" ones. President's Rule was invoked ten times between 1947 and 1967, and between 1967 and 1986 seventy times. However, in 1994 the Supreme Court ruled that the imposition of President's Rule is subject to judicial review, and that the dissolution of state governments can be overruled if the president's decision to do so "was 'malafide' or based on 'wholly irrelevant or extraneous considerations'" (Thakur 1995: 99).

[5] In the year 2000, three new Indian states—Uttaranchal, Jharkhand, and Chattisgarh—were created from portions of existing states, along linguistic and ethnic divisions.

This is not to suggest that the Constitution gives the Central government unbridled control over individual state governments; the extraordinary powers granted are just that—the Central government can invoke only those powers spelled out in the Constitution, and for only specific, limited periods of time. Ramesh Thakur writes that, as a federal governmental form, India

> is a flexible rather than a rigid federation, with the balance between the central and state governments varying to suit the changing circumstances.... Under normal conditions, the authority of state governments is coordinate with rather than subordinate to the wishes of the central government. For it is derived not from laws made by the central government, but from the very same constitution which creates and legitimates the central government (1995: 72).

In specific regard to the issue of jurisdiction over India's natural environment, the Constitution is uncharacteristically vague. In a document that is noted for the widespread inclusion of policy statements and policy direction (Tummala 1994), there is little of substance other than a "descriptive principle" which states that "The State[6] shall endeavour to protect and improve the environment and to safeguard the forests and wild life of the country" (Government of India 1996: 30). Specific concerns over the natural environment must be derived from the Enumerated Lists in the Seventh Schedule.

As is the case with other multiple-level governments, the designation of responsibilities and jurisdictions is not a clear-cut, compartmentalized one; aspects of jurisdiction over a given area often overlap. For instance, while the control of water resources—including "water supplies, irrigation and canals,

---

[6] In this context, "the 'State' includes the Government and Parliament of India and the Government of each of the States, and all local or other authorities within the territory of India or under the control of the Government of India" (Government of India 1996: 4). While the idea of "state" infers a wide range of constitutions, organizations and concepts, "government" refers more precisely to the existing governing regime.

water storage and water power" (Government of India 1996: 233)—lies with the states, the "regulation and development of inter-State rivers and river valleys," when deemed to be in the public interest, comes under central jurisdiction. Another power available to the Central government, which has raised the ire of many state authorities, has been the ability of the GoI to shift jurisdiction over items on the State List to the Concurrent List. As an example, forestry, an area originally under state control, has been shifted through amendment of the Constitution to concurrent administration.[7] As shall be seen in one case study, the "coordinate authority" between center and state governments in regard to forestry appears to have served as a means for the Central government to enter into direct collaboration with local NGOs when the corresponding state agency has refused to do so.

The interest on the part of the GoI in decentralized control of rural development, which led to the adoption of the Seventy-Third Amendment in 1992, has added another layer of political jurisdiction which often overlaps existing authorities. The amendment gives the states the power to endow village councils called *panchayats* with authority in the area of planning and implementing "schemes for economic development and social justice" which includes such matters as soil conservation, irrigation, water management and watershed development, social forestry, farm forestry and minor forest produce, drinking water, fuel and fodder, and waterways (Constitution of India 1996: 104).

In addition to their repeated efforts to decentralize the control of rural development decision-making and implementation, the GoI has created its own development bureaucracy, which

---

[7] According to some authors, forestry is an instructive example of the Central government's power to usurp the principle of federalism in the case of natural resource use. Ashwini Chhatre (1994: 2473), chronicling a bill before the Parliament that would concentrate central control over forests in India, writes that "the last two decades have witnessed an acrimonious struggle between the centre and the states for control over forest resources. The 42nd amendment [to the Indian Constitution] which shifted forests from the State to the Concurrent List, the 1980 Forest Conservation Act and its amendment in 1988 are a testimony to that."

often duplicates the responsibilities and functions of state- and local-level agencies, despite the fact that many of those functions are included on the State List (and are therefore constitutionally state matters). In the Union government there is a Ministry of Rural Areas and Employment, within which the Department of Rural Development has wide-ranging responsibilities in many subject areas. The District Rural Development Agencies (DRDAs)—which, though designated as autonomous bodies and registered under the Society Registration Act of 1860, are agencies of the Department of Rural Development and headed by District Collectors—are "the implementing agency for all the rural development programmes [of the central government] at the district level" (DRDA 1999). Other agencies of the GoI include the Council for Advancement of People's Action and Rural Technology (CAPART), created in 1986 "as a nodal agency for catalyzing and coordinating the emerging partnership between voluntary organizations and the Government for sustainable development of rural areas" (CAPART 1999); and the Department of Programme Implementation, designed specifically to monitor Central government programs, including those involved in rural development.

Although these departments were ostensibly created with the intention of working with, or coordinating, state and local level governments, their actual operations have not been viewed as such by many observers. CAPART, for instance, is seen by many state- and local-level officials as a body that regulates and constrains NGOs to a greater extent than it assists them. Similarly, the motive behind the GoI's use of the power to shift jurisdiction from state control to concurrent jurisdiction is a matter of conjecture. Some authors claim that this is a justified activity on the part of the Central government to bypass intransigent state governments and agencies, while others see the GoI's actions as "the steady pilferage by the Centre of powers that constitutionally belonged to regions" (Khilnani 1997: 51; see also Baxi 1986, Tandon 1991).

Environmental Politics in India

Contemporary concerns regarding the natural environment, pollution control and natural resource conservation are of relatively recent origin in India. Despite the longstanding problems, interest in issues of the natural environment did not become widespread until the early 1970s, during the first administration of Indira Gandhi as the Prime Minister of India. Renu Khator points to 1971 as the watershed year for the emergence of environmental issues in the country. In that year, following decisive political successes in India (both a landslide electoral victory for the Congress, and the overwhelming defeat of Pakistan in a war that resulted in the creation of Bangladesh), Mrs Gandhi attempted to expand her political influence by committing the Central government to an expansion of economic and social reforms, including those with respect to the natural environment. Khator, while recognizing international influences that accelerated the process of legitimization of environment as part of India's policy agenda (noting the Prime Minister's active participation in the 1971 United Nations Conference on Human Environment), writes that "[Mrs] Gandhi's personal commitment was the single most crucial element in the formation of India's environmental agenda" (1991: 66).

As a result of Indira Gandhi's focus on environmental issues and international pressures stemming from a number of international conferences, the GoI passed a number of environmental laws between 1974 and 1986. These include the Water (Preservation and Control of Pollution) Act (1974), the Air (Preservation and Control of Pollution) Act (1981), the Forest (Conservation) Act (1980) and, in light of the failure of existing laws to stem destruction of the environment (and, more immediately, in the wake of the Union Carbide disaster of 1984 in Bhopal), the Environment (Protection) Act (1986). Khator (1991: 78) notes that, "It was realized that several sources of environmental destruction, such as hazardous substances and acid rain, were not covered under existing laws; therefore, a comprehensive environmental law was needed. The [Act] reflected this major concern."

The passage of environmental legislation was accompanied by the creation of several government agencies, beginning with a National Committee on Environmental Planning and Coordination (NCEPC) in 1972. The NCEPC was an advisory board whose responsibilities consisted of suggesting and reviewing environmental policies and programs, to "ensure that while the country pursues its legitimate desire to develop its economy as fast as technology and resources will allow, the environment and the quality of life that depends on it is not unduly degraded" (Dwivedi 1977: 128). Until its replacement in 1980, it served primarily to increase awareness in the Indian environment and stimulated the creation of an extensive environmental bureaucratic network in the GoI and more than hundred public interest groups (Khator 1991: 88). The NCEPC's role was, however, strictly advisory; it wielded no coercive power.

In 1980, the GoI created the Department of Environment (DoE) by combining the resources of several existing government departments. The DoE was intended to play a watchdog role by conducting environmental appraisals of development projects, protecting and conserving wildlife, monitoring air and water quality, creating an environmental information system conducting and facilitating environmental research, and encouraging international cooperation. In addition, it was charged with coordinating activities between the federal, state and local governmental levels.

The DoE ultimately proved unsuccessful. According to Khator (1991: 89), the primary reason for its failure was

> the inherent conflict and vagueness that existed in the structure itself. The environmental policy...was regulatory in nature, but at the same time, the bureaucratic network that it created was non-regulatory. The Department of Environment was an advisory body with a small political base and even a smaller financial support. It was a "watchdog" who had no bark.

From its beginnings, the DoE became vulnerable to various pressures that would, in a few years, lead to its demise. These stemmed from industrial and business interests, the corrupt and rigid nature of the Indian bureaucracy (which led the DoE to greatly restrict the discretionary powers of local governments), and within the government itself. Without an independent power base, the DoE could not oversee the activities of other development-related agencies, including the Forest Department (which will be discussed below).

Despite the failure of the DoE, the GoI did not re-evaluate its own environmental policy; instead, the government encouraged policy-makers to tighten their control of the bureaucracy. One result of this approach was the creation of a new, more powerful institution with a greater degree of government control; in 1985, the DoE was reborn as the Ministry of Environment and Forestry (MoEF), an independent ministry (akin to an American cabinet-level agency) with wide regulatory capacities.

Despite the existence of a widespread bureaucracy and a large, vociferous public, the inability of the GoI's environmental bureaucracy to effectively address the country's environmental problems has consistently increased. In addition, although the Central government has made numerous attempts to decentralize and democratize the rural development policy process, there remain a number of structural, institutional, and political constraints to the effective operation of governments in India in their attempts to bring about changes in the conditions of the rural poor.

Given the scope and authority granted to the Central government through this myriad of legislation and the broad powers of pre-emption derived from the Constitution, one might expect that the GoI would utilize such far-reaching authority to direct state activities regarding the environment. Undoubtedly, it has moved decisively and rapidly to implement some of its policies, yet the actions of the Central government have been, to a large degree, ineffective. Why?

Khator (1991: 194) suggests that environmental policy in India, as in most countries (including the developed countries),[8] results from a policy deficit—that is, "the gap between policy

---

[8] For a discussion of this phenomenon in the Western context, see Vogel (1986).

objectives and policy performance." The shortcomings in Indian policy implementation occur, in part, because

> political processes in [India] are not geared toward reducing the deficit; they are instead tuned to find an equilibrium where the environmental issue can be made non-threatening. The environmental process, therefore, emphasizes non-scientific, politically feasible and reconciliatory objectives.... A strong desire to reconcile, rather than confront, has been prevalent throughout the policy process. (Khator 1991: 196)

More specifically, Khator writes that environmental policy-making is strongly influenced by, and subordinate to, developmental needs. The GoI faces the necessity of presenting a public image of achieving success in both areas; because developmental and environmental goals are often seen as inimical, and because achievements in the area of development are generally more visible and more quickly realized than environmental ones (increasing forest cover or abating water and air pollution are less immediately apparent than are more readily quantifiable economic gains), the government has adopted what Khator calls a "pretentious policy approach" to its environmental efforts. In order to present a positive image of its environmental bureaucracy (despite a meager record of actual environmental improvement), government reports emphasize activities, rather than achievements:

> Policy outcomes are measured in terms of how many networks have been established, and how many emission standards have been set up, instead of whether the quality of water or air has been improved as a result of these networks and standards. The government is able to achieve a balance between its developmental goals and environmental goals by glorifying procedures, and as long as these procedures do not directly threaten the developmental goal, the government faces no political challenge by committing itself to environmental activities (Khator 1991: 100).

As mentioned, the responsibility (and the expectation) for the implementation of development programs in India and other LDCs has generally rested with national governments. Since the perception of environmental problems in developing countries tend to be focused more on natural resource depletion and equitable use than on industrial pollution, wilderness conservation or amenity issues, which are more common in industrialized countries (Redclift 1987; Kothari 1989; Guimarães 1991; Krishna 1996), resulting policies are directed at environmental degradation or exploitation of resources and are likely to be regulatory in nature. The official attitude adopted in many (if not most) LDCs with regard to such common property resources has been to prevent their loss, and for good reason: the extraction, utilization and export of natural resources represents a major portion of the income available to many developing and industrializing states' economies (see Ascher and Healy 1990; Berkes 1989). Consequently, as Khator (1991) posits, the assumption on the part of governments is that "irrational" use of such goods (those uses regarded as incompatible with the national interest) must be prevented through government intervention; regulation is necessary to forestall uncontrolled resource appropriation by "free riders."

The role of government as regulator has been a characteristic of the Indian bureaucracy since long before Independence (as mentioned above, the highly centralized administrative apparatus of the Moghul dynasties, which had broken down completely with the arrival of the East India Company, was reconstructed by the British to suit their own administrative needs). Nehru modeled the bureaucracy after the British system, denounced by so many Indians at the time of Independence, as the best way to lead the new country into the modern world.

This paternalistic attitude, which still predominates the GoI, is exemplified in the passage of the Environment (Protection) Act in 1986. The Act was passed (at least in part) as a reaction to the tragedy that occurred at the Union Carbide plant in Bhopal two years earlier. Rather than taking new directions in environmental management, its objectives were to cover loopholes and omissions in existing laws and to strengthen the authority of the GoI.

According to Khator, the Government of India has chosen to address environmental deterioration through regulation, rather than working cooperatively with industries, individuals, and other polluters.

> The regulatory strategy offers some unique political benefits. It gives the pretense of urgency, a sense of governmental activeness, and also a means of reconciliation. The regulatory approach of India, for instance, allows the two conflicting policies—of economic growth and of environmental preservation—to coexist. The changing needs of society continue to find and modify the equilibrium between the two (Khator 1991: 206).

Policy decisions are made so as to reflect the overall values of society, such as democracy or capitalism. They must also conform to basic political imperatives such as partisan politics, parliamentarianism or federalism. It is at this point that the rural poor have their greatest access to direct participation in the policy process, through demonstrations and protests (Khator 1987: 156). As Sumi Krishna points out, the environmental debate is based less on analysis and understanding of the issue than "on the growing strength of protest movements among the people most affected by ill-planned and badly-executed development projects." Indian environmental concern is "criss-crossed with contradictions" (Krishna 1996: 35).

Several authors note the diversity of approaches to the environmental problems in India. Gadgil and Guha (1995: 63) write that the number of natural resource conflicts has increased as the Indian government has lost its legitimacy as the "authentic legatee of an all-class and genuinely mass-based national upsurge" following Independence. At the same time that the state is popularly seen as a representative of the interests of a narrow elite, the democratic system of the country "has conferred on the growing number of ecosystem people [those depending on their own local natural environments for the bulk of their material needs] and ecological refugees a modicum of political clout." The result is that the debate over the environment and the causes of its ills is couched in terms of whether development is the cause of the environmental problems or the solution.[9]

The dominant approach to environmental issues taken by the GoI, as well as a number of NGOs, is a managerial one. Adopted from Western approaches, the managerial approach is grounded in scientific assumptions about the environment: that an optimal balance of resource uses can be found that accommodates production with conservation; that modern technology is the most effective tool for utilizing and managing resources; and, therefore, that government policies and interventions to reverse adverse environmental damage are the appropriate answer. The managerial approach is thus a techno-centric rather than eco-centric approach to environmental concerns. It views environmental degradation as a result of shortcomings in the development model, yet seeks to fine-tune that model rather than reject it. Development is viewed as both the cause and the solution of environmental problems (Krishna 1996; Redclift 1987; Sachs 1993). The approach is concerned with techniques more than with policies; it reinforces and facilitates the predominant Western industrial approach to the environment.

The Policy Process: Formulation and Implementation

In India and other developing countries, the failure to adequately address development, including natural resource and

---

[9] Krishna (1996: 43) warns that "as the ideological debate has evolved, the theoretical frameworks have become idealised models which are far removed from the reality of immediate environmental problems. The tendency now is to look for universal truths from particular social situations, rather than to derive pragmatic solutions for particular problems. ...Theoretical positions on critical issues—such as the curbing of excessive consumption, the integration of ecology and economy, the democratization of community management of resources and so on—cannot substitute for practical answers to the problem of how this is to be done."

environmental problems, may stem from any one of many points in the policy-making process, as well as in the communication of policies and objectives from one level of government to another. The complex array of processes by which development decisions are formulated, transmitted, and implemented is an important consideration in understanding critical collaboration.

One of the earliest bottlenecks in the policy process often occurs at the point where the issue at hand is identified. All too often, an environmental problem is cast in terms of a single, isolated phenomenon and a corresponding solution. Perceptions of other, interrelated, components become of secondary importance as attention is focused on the most visible (see Ascher and Healy 1990). Responsibility for the solution of an issue tends to fall to a single agency with the political mandate for, or the technical knowledge of, the issue. Technical experts, having been trained to see issues primarily in terms of their specialized education, frequently dominate the process and further isolate the situation by focusing on a limited range of aspects regarding what are usually multidimensional problems. Another facet of this reductionism lies in the tendency to apply one set of solutions in a wholesale fashion, overlooking the myriad of local differences that require the adaptation of programs to individual social and natural environments for their successful implementation.

Politics also plays a role in couching such complexities in simplistic terms. In the context of developing countries, B. Bowonder (1983: 377) writes that "[w]ith elections to win, wars to fight, dams to build and hungry mouths to feed, it is hard for any politician to concentrate funds and attention to problems which are multidimensional and seemingly long term in nature." Natural resource and environmental issues are especially prone to such simplification because of the complex linkages and interactions between "the whether, how and when to convert resource endowments into capital and consumption—and for whose benefit—in the context of both positive and negative externalities" which do not lend themselves to facile solutions (Ascher and Healy 1990: 175). Ascher and Healy also point to another, more blatantly political factor favoring the adoption of simplistic solutions: many authorities opt for simple, single-objective solutions because they provide significant short-term political rewards, despite politicians' awareness that such symbolic gestures usually engender future difficulties as new problems surface or old ones defy resolution (1990: 164).[10]

The participation of the populace—those for whom development efforts are intended—in the identification and shaping of development policies has become a major issue in almost all development circles, and in the absence of organized avenues for public participation in the policy process, protest becomes an alternative means of affecting government decisions. Specifically in India, one major aspect of the political landscape is the degree to which mass protests and demonstrations have become a part of the everyday political scene and the extent to which they are tolerated by the government (although the efficacy of such popular action is not assured). Hardgrave and Kochanek attribute the stability of Indian society and the ability of the government to effectively manage such activism to three factors. The first is the strength of the political institutions that emerged from the centuries of British colonial rule. The second involves a "broad-based commitment" to democratic principles and politics "by which expanding participation can be absorbed and ordered" (Hardgrave and Kochanek 1993: 176). Finally, the authors hold that the complexity inherent to India's society is itself a source of social stability. Indian society is composed of so many cross-cutting divisions—based on class, caste, religion, languages, tribes, sects, and regional differences—that the unrest of one social group is seldom translated into a mass movement at the multiple-state or national level, and can therefore be effectively managed and controlled (Hardgrave and Kochanek 1993: 175; see also Rudolph and Rudolph 1987; Khator 1991; Mitra 1992).

[10] An illustrative example of these difficulties—both defining the issue, and the political considerations involved in simplifying the issue—is the case of the construction of a series of dams on the Narmada River and its tributaries (see Colchester 1985; Alvares and Billory 1987; Thakkar and Kulkarni 1991; Appa 1992).

Furthermore, the Indian federal system has served to isolate unrest by creating individual states according to differing linguistic and social identities.

One may argue that popular protest is such a widely accepted political activity in India that it might be thought of as having become institutionalized (Mitra 1992: 8), and that the role of indigenous NGOs in championing popular concern over issues of development and environment could be considered a logical extension. In fact, the degree to which the GoI has accepted popular protest—and NGO involvement in the same—is mentioned by Fisher (1997: 64), who writes that "In India, it is now official government policy to encourage GRSOs [grassroots support organizations] to organize the poor in order to make bureaucracies more responsive." It is clear that official GoI policy specifically calls for the participation of NGOs in the government's policy arena, even in cases where unwelcome pressure is directed at government agencies in order to change their attitudes and procedures. This is an indication that popular, and advocative participation—and by inference, collaboration—is an official policy of the GoI, although this is not necessarily translated into practice, especially at the level of subordinate governments.

Ultimately, though, it is at the local level, where environmental policy is interpreted and implemented, that a number of barriers to successful progress exist. It is here that well-meaning, coherent policies drafted at the national government may be thwarted. Mitra (1992: 22) writes that

> It is important to emphasize here that more than the high politics of Delhi, regional politics and policy process constitute a critical level for the analysis of development in India. In this sense, the states of India, seen as the regional policy environment, are the appropriate units of analysis. Generally, it is at the level of the regional process that the competing claims of development agencies are negotiated. Often, in practice, a minister of the state government, representing the overarching role of the state, is called upon to arbitrate among the competing claims of the different agencies and

to lay down the authoritative norms to act as a guide for their operation.[11]

The Local Context

At the local level—especially the district and villages—policies created at the centers of political power are subject to a distinctly different context, what Khator calls the "local policy culture" (LPC). The LPC consists of "the reflection of attitudes, beliefs and orientations of policy participants toward public policies" and provides the "boundaries" within which policies can be effected (Khator 1991: 143). The LPC in India, according to Khator (1991: 144), contains,

> several general perceptions: 1) that governmental officials are corrupt, incompetent, and vulnerable; 2) that government policies are insensitive to the needs of the general public; 3) that political institutions are designed to benefit the elite; 4) that the power of the elite is overwhelming; 5) that the public does not trust the government; and finally; 6) that policies do not change anything.

Within this context, there are several important institutions that have a significant bearing on the conduct of rural development. While there are others that certainly have an influence, the following have a direct role in the conduct of rural development in their immediate area.

Local Elites

As in any society, there are individuals at the local level whose economic and/or political power position them to exert substantial influence over the direction of development. The role

---

[11] Indeed, one author warns that NGO appeals for the Central government to intervene in attempts to overcome local rigidities may, in effect, be undermining the idea of decentralized community control of development and, instead, centralizing control of the process (Krishna 1996: 128).

played by local elites in the villages of India is generally viewed as a disruptive one. Mitra (1992: 2) writes that "local elites, seek to manipulate the allocation of resources and to influence the pace and direction of social change [primarily through] skimming off the benefits of development for their private gain." At the same time, however, Mitra views their intervention in local politics as serving an important function in rural India: "India's local elites are a crucial hinge group whose ability to incorporate newly emerging social forces into the political arena and to ease out the old style notables is the crucial determinant of the stability, legitimacy and expansion of the state" (ibid.: 12).

Krishna Tummala (1994) views the role of rural elites in India in local politics today in more limited terms. According to him, there has been an emergence of popular participation in local politics resulting, in part, from popular demands for government recognition and provision of services, which is transforming "the old pattern of clientele politics" into "demand politics" (see Frankel 1971). Thus, combined with the uneven increase of mass participation resulting from the rise of Panchayati Raj institutions and the increased rule of different (and often opposing) political parties at the central, state and sub-state levels, the influence of elites has become only one of several factors in a turbulent atmosphere of local politics.

This (often) dysfunctional climate is exacerbated by the structure of the bureaucracy. At the local level, ostensibly because of the need of the center's policy-makers to control the situation, almost all possibility for discretion, individual initiative, or imagination is stifled. Jaix (1992: 89) notes that local government institutions,

> are the creatures of the state government, and carry out only those functions and responsibilities which are *specifically* delegated to them under the state legislations. Thus the composition and functions of local government institutions may vary from one state to another in the states and ... union territories existing in the federal system of India [emphasis added].

Stripped of any possibility of autonomy the local functionary, according to Kothari (1961: 825), protects his own precarious position by resorting to procedure: "[r]ed tape becomes a technique of self-preservation, and reverence for traditional forms is matched only by attachment to strict routine and an unwholesome preoccupation with questions of accountability." Underlying this penchant for strict adherence to the rules is the pressure of local politicians and elites on the bureaucrats for favors—corruption is a way of life throughout Indian government, perhaps more so in the villages than in the urban centers.

The District Collector

At the focal point of local rural politics and relations between local government and other levels of Indian government is the district collector, almost universally acknowledged to be the single dominant actor in rural India.

The office of the district collector in India is a long-standing feature of government control and centralized power in rural development. The position was initially established by British colonial rulers in the immediate aftermath of the Sepoy Mutiny of 1857 as a means of consolidating colonial control. According to Gaikwad (1978, quoted in Jain 1985: 61),

> [t]he extremely pragmatic British evolved a system of administration most suitable for keeping the unorganized, rural population under subjugation and for exploiting the natural and human resources for the benefit of the colonial power. District administration was the strongest unit of this system. Through procedural means and the raw use of authority, district administration prevented organised resistance of the rural population to the systematic exploitation of this class by the government and other vested interests.

Nominally, the principal tasks of the collector's office are the collection of revenue and the maintenance of law and order

(Sadasivan 1980, quoted in Jain 1985: 62). The job description, though, does not convey the scope of the almost unbridled local political authority that has become the hallmark of the collector. The collector has "literally thousands of enumerated powers as head of the magistracy, revenue collection, other government departments, and urban and rural authorities. In exercising these powers, the IAS Collector engaged in a central way in the political process of the district" (Potter 1996: 222).[12]

Despite a number of studies, reports and recommendations advocating administrative reforms, the district collector remains one of the most influential and powerful representatives of the GoI at the local level, where development programs and projects are ultimately implemented. Given the extent of the control wielded by the collector in almost every detail of local administration, it is noteworthy to understand that the collector is not a functionary of the district or state. Collectors are employees of the GoI's All-India Services and are "recruited, selected and trained by the Centre which also governs their service conditions" (Tummala 1994: 126; see also Thakur 1995). Thus, the potential for contravening state decisions with those of the Central government is a constant factor. While the collector is formally under the control of the individual state to which s/he is assigned, the officer is still governed by the rules of the Central government which "could not be altered or interpreted to his disadvantage except by, or with the approval of, that government; and he could appeal to the central government when he considered that his service rights had been infringed by an order of the state government" (Potter 1996: 163–4).

Given their broad mandate, collectors have the ability to act personally on a particular issue or concern of personal interest.

---

[12] Since Independence, the powers of the district collector have expanded to include "maintenance of law and order, control of crime and administration of justice; revenue administration including land administration and collection of land revenue and other public dues; control, regulations [sic] and distribution of food suppliers; arrangements for holding elections; administration of local bodies both urban and rural; welfare activities; emergencies and natural calamities and other such matters such as small savings scheme [sic] and contributions to public loans" (Administrative Reforms Commission 1967, quoted in Jain 1985).

The collector has sufficient discretion to apply considerable influence to a special interest in the district, and is "continually nursing the base of support needed to assist, or at least not thwart, the pursuit of what he or she wanted done in the district on behalf of the government" (Potter 1996: 225).

The collector represents the tendency of the Central government to retain a degree of political control at the local level despite considerable, repeated efforts on the part of the GoI to decentralize the development process and to empower the rural poor in many areas of local decision-making.

Panchayats

In its efforts to overcome the ineffectiveness of local policy implementation, and to make local government more representative, the GoI has attempted on a number of occasions to revitalize an ancient form of local government: the institution of Panchayati Raj. The system, at the village level, consists of a council of elders that, for centuries pre-dating the British presence in India, constituted a form of democratic village self-government. The rule of panchayats declined during the British reign; by mid-nineteenth century, they had no effective role in rural India "as a result of improved communications, increased mobility, and a centralized administration that emphasized the individual in society and not the elders of the village" (Hardgrave and Kochanek 1993: 112).

The primacy of villages and village independence was at the center of Mahatma Gandhi's view of an independent India. Despite the misgivings of a number of its founders, a provision for the creation of panchayats (Article 40) was included in the Indian Constitution's Directive Principles of State Policy (Part IV).

The first attempt at creating a panchayati raj was introduced in 1959 as a result of the findings of the Balwantrai Mehta Commission Report of 1957. At its zenith, more than 98 percent of India's villages and 97 percent of the rural population were involved in the system. While the intent was to devolve the responsibility for all planning and development work to local jurisdictions, the village-level panchayats "were made subordinate units of the government to implement its programs at the lowest levels, and not the agencies of self-government or Gram Swaraj as Gandhi had advocated" (Singh 1994: 819). However, by the mid-1960s, it was clear that expectations that the creation of panchayati raj would lead to a major political and administrative restructuring would not be fulfilled. In most areas, panchayati raj languished.

Since then, several efforts have been made to revive and strengthen a system of panchayats in order to enable local participation in local government, rural development, and economic planning. An effort was made by Prime Minister Rajiv Gandhi to bypass state governments so as to more readily reach villages (see Krishna 1996). The adoption (in 1992) of the 73rd Amendment to the Indian Constitution and the passage of the Panchayati Raj Act (1993) have, to a degree, actually resulted in the revitalization of some panchayats. However, their actual functioning has been uneven from one state to another, in part because the individual states govern the composition and authority of their panchayats as well as the financial arrangements essential to their operation (Kannan 1993; Kumar 1995; Sastry 1995). Singh (1994: 821–22) sees the failure of the institution's performance as mainly "the political repercussions of electoral competition between state and [panchayati raj] representatives.... State level leaders [began] to see these local institutions as rival" and reacted in several ways. Some states have withdrawn important functions from their panchayats; others, such as Tamil Nadu, have repeatedly postponed elections, causing the central Ministry of Rural Development to withhold rural development funds from the state (Kumar 1995: 2235). The Karnataka Panchayat Act of 1983, designed to give increased power to local panchayats, was overturned when a rival political party came into state control (Krishna 1996).

In several states, panchayats have been co-opted by the ruling political parties to serve their own political agendas. For example, panchayats in West Bengal, sometimes called "red panchayats," are widely considered to be pillars of the ruling

Communist Party's (CPI-M) rural power base (Bhattacharya 1993; Kohli 1984). And in the state of Kerala, according to K. R. Sastry (1995: 1909), state legislation purported to strengthen that state's panchayats was actually

> strategically tailored to concentrate power in the hands of government and officials under the pretext of decentralization of power .... In spite of all the protests, [the Act] as passed by the legislature is an instrument aimed at maintaining centralized powers with the state government."[13]

As with other GoI initiatives to decentralize control of the development process, the creation of village panchayats has not automatically resulted in improved delivery of government efforts. In many areas, the District Collector is an official or *de facto* member of panchayat institutions; in some cases, s/he has the capacity to inspect the activities of local panchayats, and can even cancel or suspend the implementation of their decisions (Tummala 1994: 130).

This largely failed attempt to decentralize rural development policy-making has been one reason for the growing focus on the role of NGOs, both as an alternative to government in delivering rural development interventions, and as intermediaries between rural populations and government agencies.

Conclusion

Efforts at development in many developing countries continue to be dominated by their governments, despite repeated evidence that these have been, to a large degree, ineffective. The attributes of the Indian government, as well as the institutional and political contexts within which development interventions are attempted, are important to the process of rural development in this country. However, there remains much to be

---

[13] See L.C. Jain (1985) for an extensive study of panchayats and their role in development.

desired in regard to the history of government-led development efforts. As explained, the GoI has made a number of attempts to include outside actors in the development process, although proclamations from the center generally have not translated into action at the local level.

As in many LDCs, the various levels of Indian government are subject to a number of institutional, bureaucratic, and procedural impediments to the effective formulation and delivery of rural development efforts. Despite this knowledge, there should be no expectation that governments in India or other countries are going to relinquish their role as primary actors in such efforts.

Given the shortcomings of government-dominated efforts at rural development, scholars and practitioners have repeatedly searched for new, more effective approaches and actors; one of the most promising has been the voluntary non-governmental sector. The purpose of the following chapter is to identify non-governmental organizations (NGOs)—what they are (and are not), how they differ from government, and how their role in rural development has evolved. This is not to suggest that government policies and activities should be circumvented or abandoned; rather, the review serves as a preface to an investigation of the concept of critical collaboration, a growing form of cooperative effort involving NGOs as partners in the rural development policy process. The concept of critical collaboration will be more fully discussed in Chapter 4.

Chapter 3

Non-Governmental Organizations

> Modern voluntarism ... aims at achieving development and social justice rather than relief and welfare. Modern voluntarism strives to change the social, economic and political position of the poor, the deprived, the oppressed, and the weak. In the final analysis, therefore, it aims at the redistribution of power, wealth, and status.
>
> Anil Bhatt (1985)

In the previous chapter, the attributes of the Indian government and its bureaucracies germane to this study were outlined and briefly discussed. This chapter reviews the nature of NGOs in developing countries and identifies dimensions of those NGOs—both in general and more specifically in India—which are important in fostering the creation of collaborations with governments. This is done because an understanding of both NGOs and GOs in the context of rural development is fundamental to an inquiry into the possibility of their working together as critically collaborative partners.

Much as the previous chapter highlighted some of the features of government attitudes toward development, this chapter begins with a discussion on the nature of NGOs in general, viewing some of the characteristics that are attributed to NGOs as a societal sector. This is followed by a brief review of some of the critiques of those presumed advantages of NGOs. The subsequent section looks at the historical functions of non-governmental organizations and their more recent activities. A discussion of the issue of NGO accountability and autonomy (which, it is argued, are issues central to the creation of critically collaborative relationships) follows. The final section consists of a brief discussion of the characteristics of Indian NGOs in general.

What are NGOs?

An entire sector of social activity in developing countries is increasingly recognized by governments and intellectuals, international organizations, and aid donors as a promising vehicle for addressing development problems, including concerns over the natural environmental and natural resource use. Collectively termed non-governmental organizations (NGOs), these voluntary non-profit institutions have a long history of humanitarian activities worldwide (in both developed and underdeveloped countries). It is only recently, however, that the importance of the roles they play in society, and the possibility of extending their activities into other aspects of social change and democratization, has been recognized.

It has become almost a truism in the eyes of many that NGOs are more successful than governmental bodies in attaining the goals of rural development, in terms of implementing projects and programs (Clark 1991, 1992; Livernash 1992; Edwards and Hulme 1996b; Mencher 1999). In addition, NGOs are increasingly considered important in bringing about other aspects of rural social change—serving as agents in the creation of (or strengthening of existing) civil societies and good governance,[1] and otherwise fostering democratic and alternative approaches to mainstream development policy and practice. This is a significant departure from the program/project approach to

[1] For a discussion of the controversies surrounding the concepts of good governance and the new public management, see Turner and Hulme (1997).

development that prevails among Western governments and donor agencies, as well as many LDCs.

Such laudatory assessments of NGOs' abilities are currently subject to a good deal of skepticism on the part of a number of scholars, even though they are in vogue with many practitioners and donors. Whether NGOs writ large can actually accomplish what has been accredited to them is subject to closer scrutiny. In fact, what NGOs actually are and what makes them distinct from other parts of society, what they do, what their roles in society are, and what the potential of the NGO sector is, are all the subject of continuing scholarly debate.

The concept of non-governmental organizations is not a readily tenable one. The nomenclature is, at first glance, a negative one: such an appellation points to what NGOs are not, yet says nothing about what they are. While some see the differentiation as a negative one (Anthony Judge (1995: 178) calls the term "a manifestation of organizational apartheid—reminiscent of the 'nonwhite' label so frequent in racist societies"), others view it as a conscious effort on the part of NGOs to differentiate their activities from those of governments (Frantz 1987).

Given that they are not (for the most part) creations of, nor are they directly subject to, governments, how can one best understand what NGOs are? In what area of society are NGOs, as broadly conceived, situated? The staggering number and variety of NGOs can make any kind of categorization daunting, if not meaningless. Some authors, such as Marc Nerfin, argue that they can be directly situated neither within the governmental nor market sectors (neither prince nor merchant);[2] rather, he writes, NGOs are located firmly in a "third system":

> Contrasting with governmental power—the Prince—and economic power—the Merchant—there is an immediate and autonomous power, sometimes patent, always latent:

[2] Hulme and Edwards (1997: 6) consider the attributes of NGOs—"[t]heir relationship with the 'people' is seen as giving them greater public legitimacy than government while their managerial features are seen as permitting private sector levels of cost control and efficiency"—to be so manifestly evident that some donors see NGOs as both prince *and* merchant.

people's power. Some among the people develop an awareness of this, associate and act with others and thus become citizens.[3] Citizens and their associations, when they do not seek either governmental or economic power, constitute the third system. Contributing to make patent what is latent, the third system is one expression of the autonomous power of the people (Nerfin 1986: 4–5; see also Korten 1990, Korten and Quizon 1995).[4]

There are several possible arguments for viewing NGOs as a distinct part of society. Frantz (1987) sees NGOs as unconnected to the private sector; although they are not public bodies in the legal sense, the way in which they interact with the population clearly places them in the public domain. Ernesto Garilao (1987) holds that NGOs should be recognized as an unambiguously singular set of actors because of their function in the society. He writes (1987: 116) of the growth of the NGO sector as the foundation of "a new service industry— the social development industry." This idea, of NGOs as an industry or social sector is, of course, an artificial designation. As Sheth and Sethi point out, the idea of non-governmentalism in India grew out of the colonial government's displeasure with various activities of the traditional organizations which were an essentially organic part of Indian society. As the British sought to control the activities of various indigenous groups which, in their perception, became involved in anti-colonial

---

[3] The notion of "citizen" in this context includes more than a person domiciled in a given geographical territory. According to MarcNerfin, such individuals are those "whose situation in society and/or some personal—intellectual or spiritual—reason makes them anxious to improve their lives, individually or collectively, and that of others.... A worker usually remains a worker, and his/her reasons to be active in a trade union are part and parcel of her/his social existence.... But not all workers ... become citizens" (1986: 6–7).

[4] In contrast, Norman Uphoff (1996) writes that, given the recent and growing decline in both fiscal support and public credibility of governmental bodies, and the growing ideological support for market-based institutions, NGOs are mistakenly described as composing a third sector. More accurately, in his view, a putative third societal sector is composed of peoples' associations and membership organizations, which engage in voluntary action and collective self-help.

movements, "the colonial state became a reference for defining [voluntary organizations'] scope of activities and their legal activity. The idea of voluntary work began to be understood as non-state or non-governmental activities" (Sheth and Sethi 1991: 51).

As Uphoff (1996) notes, an important characteristic of NGOs is that they are not what Carroll (1992) and Fowler (1991) call membership organizations—that is, they are not composed of the very people on whose behalf they function (again, a negation). Rather, most indigenous development NGOs are staffed with middle-class operatives whose backgrounds and educations (and often cultural and ethnic bases) are, in many cases, quite unlike those whom they serve. In this way, they differ from grassroots organizations (GROs), which Fisher (1997: 6) defines as "locally based membership organizations that work to develop their own communities."[5] Despite the criticism that this detachment limits accountability on the part of NGOs toward their rural constituents (compared to the level of accountability among grassroots membership organizations) and that direct government-GRO relationships might result in more democratic development, a number of authors warn that GROs are themselves subject to their own shortcomings and limitations, and that NGOs have an important role in strengthening the effectiveness of GROs as well as acting as

---

NGOs, because of the nature of relationships they form with those they serve, are more accurately situated in the private sector, despite the fact that they do not operate for profit. For Uphoff, NGOs are "service organizations" that deal with their "clientele" much as do private enterprises; they and their clients and beneficiaries form a business-like relationship, and NGOs are essentially unaccountable to their public: their clients "did not create the organization they are dealing with and cannot hold [the NGO] accountable for its actions in the same direct way that members can" (p. 25). This view does not, however, explain why (as mentioned above) NGOs are so widely viewed as a distinctive sector of society, or why they are seen as autonomous of control by either government or business (even though they generally receive both funding and authority to operate from organizations in those sectors).

[5] While GROs are certainly, in the majority of cases, non-governmental, Fisher (1997: 5) notes that the term NGO has "numerous culturally specific meanings;" she points out that the term NGO is used by many (as is the case here) to refer specifically to "intermediary or grassroots support organizations [GRSOs]."

intermediaries between government and the rural poor (Brown and Korten 1989; Brown 1991; Clark 1991; Fisher 1997).

However insufficient a descriptor, the term non-governmental organization is almost ubiquitously employed to categorize a vast array of associations that perform a large number of diverse purposes. It is not surprising that the complexities and variety of attributes of such institutions do not easily reflect such a reductionist designation.[6] Even within the field of Third World socio-economic development, in which the term is used to describe a group of organizations whose stated purpose is to provide development assistance, there exists a wide variety of institutions—from the minuscule, locally based, loosely established voluntary organization operating in one or two villages in a developing country to the large transnational organizations headquartered in the North, with offices worldwide and paid staffs of hundreds—and a panoply of mixed types in between which vary in a number of important ways. This diversity is characteristic of Indian NGOs, as will be discussed below.

For the purposes of this study, the term "NGO" will refer to indigenous, southern, development-oriented NGOs, formally constituted associations "established with the explicit objective of achieving certain ends ... and formulating rules to govern the relations among members of the organization and the duties of each member" (Blau and Scott 1970, quoted in Frantz 1987: 122). They are not composed of the rural poor, who are their beneficiaries; many NGOs include well-educated urban middle-class individuals who have chosen, for a number of reasons, to help the rural poor through the non-profit venue. The shared and universal strength of NGOs is their desire to "associate and to pursue common objectives through voluntary action, independently of government and without the expectation of economic profit" (Esman 1991: 102–3).

---

[6] According to several authors, the concept includes a widely diverse range of associations. Judge (1995) categorizes no fewer than thirty distinct forms of NGOs; he argues that even crime syndicates can be justifiably termed as such.

The Debate over NGO Comparative Advantages

NGOs have, in recent years, been considered by scholars, donors, and many outside governments as a positive alternative to government-led approaches to development. They are no longer viewed solely as implementors of development programs, but as a vehicle that may usher in a new overall approach to the problems of development. Jessica Vivian (1994: 169) writes that many Western governments and funding agencies view NGOs (as a whole) as a "magic bullet" that can be fired in any direction and still find its target and deliver positive results, despite the fact that NGOs generally leave little evidence to support the results of their activities. They are often expected to be able to come up with quick, simple answers to complex, long-standing problems that have long eluded attempts by institutions with greater resources and power, many of which have themselves sought easy, quick highly visible answers (see Ascher and Healy 1990).

Whether these claims are true (and few have been rigorously substantiated), NGOs are increasingly compared to LDC bureaucracies by Northern media and aid-givers and described as "virtuous Davids fighting the Goliaths of famine, hostile climate, government inequity, slavery, and oppression" (Clark 1991: 45). NGOs have gained the favor of Northern intellectuals as well. There has been a growing call for collaboration between NGOs and GOs by academics and practitioners (Carroll 1992; Clark 1991; Esman 1991; Farrington and Biggs 1990; Korten 1987) to enhance the impact of government poverty-reduction programs, and to democratize the development process (see Bebbington and Farrington 1993).

Of the qualities most often attributed to NGOs, perhaps the most commonly held is related to their autonomy. It is widely held that freedom from government constraints gives NGOs greater latitude to attempt untried approaches to development, to adapt new techniques to unique situations, and to combine known, successful practices with untested ones. This is attributed to the belief that the organizational structures of most NGOs are informal rather than bureaucratic or hierarchical, and that NGOs pay greater attention to results than to regulations and the protection of their own political power (Mars 1992).

A number of other positive attributes have been enumerated by the supporters of NGOs regarding their ability to outperform governments. Judith Tendler (1982: 6) notes several truisms of NGO activities: NGOs succeed in reaching the poor; the poor participate in NGO projects; there is greater emphasis placed by NGOs on teaching people how to gain better control on their lives than on the attainment of specific tasks and goals; NGO programs/projects can be (and are) more flexible and experimental than official ones because they are smaller in scale; their project costs tend to be lower than government-led ones; and, as private organizations, NGOs have "a special ability to work with and strengthen local institutions."

NGO interventions are also perceived as more effective and sustainable over long term than the government programs (Robinson 1992). NGO workers are characterized as inherently highly motivated, committed to their work, and providing a more continuous, long-term presence in a community than frequently transferred government employees (Leach 1988; Clark 1991). Specifically in the area of the natural environment, according to Jaix (1992), NGOs have a more accurate and practical knowledge of local environmental problems than would be possible for government organizations to attain. Therefore, the use of NGOs to monitor environmental quality at the grassroots level is crucial.

Are the accolades valid? Are NGOs actually endowed with these superior qualities? This attitude prevails among many NGOs themselves, and a number of donor agencies reinforce these claims (see Cameron and Cocking 1991). There is, however, a paucity of concrete evidence to back up such claims. According to John Clark (1991: 53),

> there is little objective reporting of NGO projects. Northern NGOs' own writing generally concentrates on the success stories and, being aimed largely at their supporting public, serve a propaganda service. ... Most contributors have an

implicit faith in the "NGO approach" which they don't want rocked. After all, one doesn't scrutinize magic too closely, otherwise it loses its charm.

Many of the positive characteristics attributed to Southern NGOs have been challenged in recent years by investigators as NGOs emerge from relative obscurity (see Bowden 1990; Brodhead and Herbert-Copley 1988; Clark 1995; Crombrugghe et al. 1985; Kottack 1985; Livernash 1992; Riddell and Robinson 1995; Tendler 1982; Vivian 1994). In addition to the critiques of NGOs' supposed comparative advantages, the viability of a long-held conceptual distinction between top-down and bottom-up (or grassroots) approaches to rural development is being challenged. Top-down approaches (typical of government-led efforts) are planned and implemented under the auspices and direction of the central governmental agencies and are viewed as overly bureaucratic and applied in a blueprint—a "one size fits all"—approach that cannot be tailored to local social, political, economic or environmental nuances. On the other hand, bottom-up approaches (generally originating both in design and implementation at the individual village-level), which are usually attributed to NGO-led efforts, are thought to be both flexible and innovative, and tailored to the unique situation at hand. While the contrasts (and the results) are held to be readily discernible and demonstrable, Uphoff (1988) writes that, in reality, distinctions between the approaches cannot be easily made. He states that while there are differences in the philosophies that underlie the approaches, it is often the case that top-down initiatives are often necessary for the activation and maintenance of bottom-up efforts. Uphoff argues in favor of strategies that transcend the two approaches.[7]

---

[7] Uphoff subsequently warns against the tendency to make similar clear-cut distinctions between two assumed fallacies underlying the top-down vs. bottom-up divide: the paternalistic fallacy, "the belief that planners, technicians and experts possess all the knowledge, wisdom, and virtue needed to achieve development, and that the poor should be responsive and grateful beneficiaries" and the opposing (and equally doctrinaire) populist fallacy, "that the poor themselves possess all that is needed for their own advancement—that they can do entirely

Considering the unanswered critiques and the lack of hard evidence of NGO claims, it would appear that the rush to support the activities of NGOs should be tempered with some caution on the part of their supporters and funders. Even so, the rosy acclamations have engendered a steady increase of financial and other resource transfers to Southern NGOs by both their own governments and bilateral and multilateral sources, including the bypassing of the more usual path of funding through government channels or Northern NGOs (Clark 1993; Bebbington and Riddell 1997). This, in part, has led to a proliferation of NGOs in a number of LDCs (Fowler 1991; Constantino-David 1992), including the countries of South Asia (Edwards and Hulme 1996a).

The growth in numbers of NGOs and the expansion of their activities, as well as the widespread criticism of governments as the central actors in rural development, provide a basis for an inquiry into the possibility of the two working collaboratively. Before considering the possibilities of new and expanded roles for NGOs, it would be instructive to understand the nature of their changing roles through the recent past.

The Evolution of NGO Functions

The work of private not-for-profit organizations in leading and facilitating efforts to improve the lives of the poor in developing countries is not a recent phenomenon. However, development NGOs have undergone a historical evolution with regard to the role they play in LDC society, both in their functions and in their status in relation to the government.

Historically, NGOs operated outside the sphere of government activities, focusing especially on the provision of disaster relief services, welfare activities, and charity work (Leach 1988: 86). Sheth and Sethi (1991: 50) write that

---

without bureaucrats and technocrats. ... While there are some impressive self-help examples and enclaves, those regional and national programs that benefit the poor on a significant scale have been concurrent mutual endeavors from above and below" (Uphoff 1988: 48).

history bears a continuous testimony to non-state efforts and initiatives toward building structures of socio-economic security by the people for themselves [including NGOs]. The state did not constitute a frame of reference for these activities; the traditional mode of organising self-help and philanthropy was essentially societal and not estatist in nature.

The predominant mode of activity is what Korten (1987, 1990) refers to as "first generation" strategies, in which NGOs responded directly to the immediate, short-term alleviation of visible needs. Efforts were directed at the amelioration of the symptoms of poverty and underdevelopment, not the causes.

NGO orientation toward development activities shifted, beginning in the 1960s, as they began to question the adequacy and appropriateness of existing state welfare activities and expressed the need to address additional, more fundamental issues of development—especially why such problems exist—in addition to (but not replacing) their earlier welfare role. This entailed a focus on understanding and confronting the causes of the problems of a local community; in addition to remedying the immediate, visible effects of underdevelopment (hunger, disease, and homelessness, among others), mobilization of local populations to address the underlying causes as well as the symptoms therein was necessary. The role for an NGO in what Korten (1990: 119) refers to as "second generation" strategies is that of an "outside change agent" which "helps the community realize its potentials" through its own interventions and guidance. According to Bebbington and Farrington, who outline a similar evolutionary path for NGOs, the second generation institutions, many of which were informed by the ideology of liberation theology and the works of contemporary authors such as Paulo Freire (1987, 1989) were, in addition, "critical of government, and most [avoided] any contact with it" (Bebbington and Farrington 1993: 202).

The next stage in the evolution of NGO thinking, what Korten (1987, 1990) refers to as the "third generation," is based on a more inclusive view of what causes underdevelopment. It is a strategy that takes into account outside forces that affect the village and local development efforts and is aimed at seeking solutions for local problems in the context of those outside policies, decisions, and constraints. This is an orientation that leads NGOs to work toward the creation of a political and institutional environment that allows the changes needed to address local problems.[8]

Making the change from a service provider to that of an effective agent of social and political change requires a basic shift in the NGO's orientation to its surroundings. An organization unused to such activities must undergo a learning experience in order to develop what Korten (1986) calls "strategic competence".[9] In addition, NGOs must create, strengthen and maintain connections with other sources of technical and managerial knowledge whose expertise they can use to strengthen their own position vis-a-vis GOs (Clark 1991; Korten 1986).

Coupled with the desire to make fundamental changes in policy, the NGO must be aware of what it is capable of achieving in its new role, and the degree to which it can take on new roles and responsibilities without threatening its own well-being. The shift to participation as a strategic actor is not simple, and requires that an NGO make a conscious assessment of its current capabilities and seriously consider the extent to which it can undertake the necessary institutional changes without damaging its well-being. Many NGOs lack the capacity and/or the commitment to remain engaged in what is often a long-term process spanning ten or twenty years. As Sen (1987: 63) points out, many NGOs—large as well as smaller NGOs—do not consider the effects of such a change. Most NGOs are

[8] Korten ultimately posits a fourth theoretical stage, a hypothetical extension of the previous. He finds the third approach, one that looks for change in individual local and national institutions, wanting in that it "requires countless replications in millions of communities, all within a basically hostile political and institutional context" (1990: 123). He calls for a world-wide campaign aimed at decentralized development action: nothing less than a "global people's development movement" centered on a "people-centered development vision" (1990: 124).

[9] For instance, one of government's political central priorities is the maintenance of its influence over local people, and one of its greatest fears is the loss of legitimacy and local support. "[T]his is a source of tension which NGOs must effectively deal with, or NGOs will continue to face limits on the nature and scope of their operations at all levels" (Riker 1995b: 112).

principally action-oriented, and find the time and effort for self-reflection to be of a lower priority than their other activities.

John Clark (1995) refers to the shift in the types of activities conducted by NGOs as they progress through this evolution as a progression, from a "supply side" approach to local development (in which organizations focus on implementing services or projects, and may work separately from, or as adjuncts to or contractors for the government) to a "demand side" orientation. In the "demand side" approach, Clark (ibid.: 593) writes, an NGO's purpose is to "[help] communities articulate their preferences and concerns so as to become active participants in the development process." At this level NGOs assume a number of roles. They assist citizens by educating them about government activities, and by helping to form citizen groups by which the poor can voice their concerns and help protect themselves against reprisals. They use the tools of advocacy and political influence to secure government accountability toward the powerless; and they create opportunities for government officials to work with, and learn from, the people for whom they work.

The idea of a demand side approach to development by NGOs implies that the predominant approach to development is ineffective and inappropriate, and that the goal of an NGO which adopts the approach is to actively change the existing balance of political, social, and/or financial power. Yet such a wide-ranging undertaking requires tools and capacities which may be beyond the reach of many, if not most, rural development NGOs:

> The moving beyond the development "supply-side" to "demand-side" activities requires that NGOs develop new skills, partnerships and ways of working in order to help communities articulate their concerns and preferences, to manoeuvre into a negotiating position with official bodies, and to mix technical operational skills with "information age" communication, advocacy and networking skills (Clark 1997: 45).

As mentioned above, NGOs in many LDCs are increasingly making conscious decisions to take on new roles and responsibilities in the name of improving both the delivery and content of development services and in a growing number of cases, the very nature of development. At the same time, expectations of what NGOs can accomplish are being revised within a number of circles. The extent to which NGOs can engender fundamental reshaping of the political and social landscape of rural areas is under investigation by governments, as well as donors and business.

An important new role envisioned for many NGOs today, and one that may well be beyond the capacity of many, stems from the widespread view that NGOs enjoy closer relationships with their rural constituents and, as a result, are more likely to engender their trust and participation than do government agencies. In the face of a dominant government that cannot help but overlook the diversities between hundreds or thousands of individual communities, the creation and strengthening of rural institutions that can oppose the homogenizing tendencies of the state and demand accountability on the part of bureaucracies is a fundamental necessity (Nandy 1989). This is the task of creating, rebuilding, or reshaping rural institutions through the strengthening of civil society.

The use of NGOs as intermediaries between the state and a largely unorganized rural society is a popular idea. Both official and private funding for rural development NGOs to operate as creators and facilitators of democratic institutions has burgeoned in recent years. However, is such an active, and even a leading, role in creating and supporting institutions of democratic government a realistic possibility for NGOs? In those societies with authoritarian governments whose attitudes toward NGOs are hostile or defensive, it is difficult to envisage such a role for NGOs (unless, perhaps, it is performed in a subversive manner). In more benign or cooperative situations, the enormity and complexity of such a task, combined with the limited willingness, ability, or capacity of NGOs to take on such a role, may simply make such activities untenable. Only a handful of NGOs in any country may be able to conduct such work. Furthermore, the idea that NGOs can create civil society and democracy relies on a simplistic notion that these are goals that can be attained through programs and projects, rather than through fundamental and continuing processes of societal and political change (Biggs and Neame 1996: 45). Yet this is the role envisioned for NGOs by some donors and governments.

In light of the praise for indigenous development NGOs as participants in the rural development process, there have been efforts to identify an appropriate role within the existing framework of development for them to operate. Although, as has been mentioned, the GoI has made the issue of NGO participation in the country's development a matter of official policy, joint NGO-GO alliances have not become a regular feature of the country's development landscape. Many GOs are distrustful of NGOs, and *vice versa*. While their freedom from government control is generally viewed as a positive attribute of the NGO sector, it also means that governments have limited knowledge and expectation of their NGOs' activities and attributes; this adds to an atmosphere of isolation between the two. The tension between these two concepts—autonomy and accountability—and their significance for critical collaboration is an important consideration.

Issues of Autonomy and Accountability

As organizations that exist and operate within a complex and dynamic web of social, economic, cultural, and political surroundings, it is nonsensical to suggest that NGOs can ever operate in complete isolation or disdain of other actors involved in the development process. The very fact that NGOs aspire to improve the livelihoods of the rural poor means that they have committed themselves, in some manner, to perform activities on the behalf of others; their ability to accomplish what is expected and promised is fundamental and necessary to their relationships with other organizations, as well as the rural poor.[10] Such organizations are not free of critique, expectations, or input from those whom they serve, receive legitimization or funding, and/or otherwise interact.

One factor important for any organization intended to play an important role in shaping the nature of society and individuals—and especially those bodies entrusted with the funds and political power intended to be used to improve the conditions of the rural poor—is accountability, "the means by which individuals and organizations report to a recognized authority (or authorities) and are held responsible for their actions" (Edwards and Hulme 1996b: 8). Those who fund, regulate, and work with NGOs always have expectations for certain activities, behaviors, and outcomes on the part of the NGO; an NGO's continuing existence often depends on its fulfillment of such expectations. As with any organization, the degree to which an NGO is required to perform in accordance with such outside expectations and/or requirements necessarily places restrictions on the organization's freedom and discretion, features that many NGOs jealously strive to maintain. Accountability is basic to virtually all sustained relationships.

As organizations that interact with a number of other institutions, groups and individuals in a number of different ways, NGOs have multiple accountabilities: to their own, as well as outside donor governments; to other funding agencies; to the rural poor for which they work; to other NGOs with which they associate; and, as Tandon (1996) points out, NGOs must be accountable to their own internal governance.[11] According

---

[10] This point challenges Uphoff's (1996: 25) assertion that the NGO-grassroots relationship is fundamentally a "take it or leave it" one in the sense that NGOs are not accountable to their grassroots constituents the same degree that organizations created by those constituents are. This is a basis upon which he makes his claim that NGOs should not be considered a third sector (see footnote 25).

[11] Tandon (1996: 55) writes that "governance implies addressing the issues of NGO vision, mission, and strategy; it focuses on future directions and long-term strategic considerations; it addresses the issues of policy in relation to internal programming, staffing, and resources; it defines norms and values that are the basis of institutional functioning; it includes obligations entailed in fulfilling

to Hulme and Edwards (1997: 142), the nature of those accountabilities can take several forms: financial, legal, institutional, contractual, and relational (see also Avina 1993). The presence of such an array of obligations and responsibilities raises several questions: Is there one organization or constituency to whom the NGO is first and foremost accountable? What forms of accountability (if any) are more readily or reasonably attainable—or measurable? And, are accountabilities to certain groups and individuals more important to an NGO's goals than others?

It is difficult to reconcile the concept of accountability with another, seemingly contradictory idea: autonomy, "the freedom to make decisions with the optimal degree of discretion" (Fisher 1997: 77). As ideal types, autonomy and accountability are completely incompatible ideas. Practically, though, absolute autonomy is an impossibility: organizations cannot, in reality, perform public activities in an operational vacuum. At the other extreme, complete accountability to all involved individuals and organizations would make it almost impossible for an NGO to do more than conform to outside demands. The extremes of complete autonomy and rigid accountability must be reconciled in some fashion if NGOs are to be able to maintain their unique organizational characteristics while, at the same time, satisfying the demands of others (see Riker 1995a; Hulme and Edwards 1997).

Is such a balance achievable? Opinions vary. There is widespread concern that NGO accountability is directed primarily "upward," toward the state and other funders, to the point that accountability to other constituents is neglected. The power of donor organizations and governments, because of the degree to which NGOs are dependent on them for financial assistance, is of such magnitude that an NGO's accountability may be skewed in that direction, while its responsibilities to the rural poor—and to its own mission to work in the interests of its rural beneficiaries, rather than its benefactors—may be

---

statutory requirements applicable to NGOs; and it focuses on defining the external positions that are consistent with the overall thrust of NGOs as institutions in civil society."

weakened. It is also important to consider that the influence wielded by those who control finances may serve to quell the voices of those NGOs whose views and practices differ from—or challenge—those of governments and donors. Accountability between and among NGOs may also be less likely to occur as they vie with each other for funds (see Edwards and Hulme 1996b).

To whom, then, should NGOs be accountable? Several authors argue that the NGO accountability should be redirected "downward"—that a greater degree of accountability to rural constituents and grassroots organizations (GROs) is vital to democratic and participatory development. In this regard, Julie Fisher states what appears to be a practical impossibility: that NGO autonomy "should be confined to what might be called upward relationships....[NGOs] and their networks should be autonomous from governments and donors." Though this may be a desirable position for any number of NGOs, their activities frequently depend upon the favor of their governments, as well as other donors. Edwards and Hulme (1996b: 12) take a more practical view: because it is difficult to objectively measure NGO performance, and because all NGOs have multiple accountabilities, issues of accountability between and among actors must be negotiated.

There is a widespread consensus among governments and funding organizations that for NGOs, as well as governments, accountability is an important organizational activity. However, little organized study of NGO accountability has been done and, in general, both public and private sector organizations seek to avoid accountability of their own activities and finances (Edwards and Hulme 1996b; Fox 1992). This has been true in the Indian context, where there is ongoing debate as to the merits of the institutionalization of mechanisms by which NGOs would be required to report their activities to the GoI. Can autonomy and accountability coexist? Although apparently inimical, it appears that the schism between the two issues must, to some degree, be bridged if both GOs and NGOs are to agree to work together as separate, but cooperating partners. This will be discussed in Chapter 6.

Although there are many commonalities to the sector as a whole, individual NGOs perform in a wide array of political and social contexts, which affect the ways that they operate. The discussion now turns to a brief review of some of characteristics of NGOs in India and the states in which they work.

Indian NGOs

India is home to a large number of NGOs. Estimates of the total number vary, from the thousands to the tens and hundreds of thousands, depending on the definition of the term (Robinson 1991). The number and concentration of NGOs varies by states and regions, with the greatest aggregation of Indian NGOs existing in the southern states, stemming in part, according to Robinson (1992), from a long history of Christian social welfare and humanitarian efforts in the region.

As elsewhere, Indian NGOs involved in rural development exhibit a wide range of attributes. In general, though, the typical Indian NGO is a small agency, concentrating on the delivery of specific services and projects to a few villages within a limited geographical area, with a handful of operatives, and limited funds and other resources. As in other countries, there is a relatively small proportion of larger, well-endowed organizations which work in several states with large staffs and significantly greater resources.

Despite their diversity, one widely perceived similarity among Indian NGOs (and of governments as well) is their focus on poverty alleviation and the empowerment of the poor (Farrington and Lewis 1993). How that is to be achieved, though, is a point of departure. According to Riddell and Robinson (1995), there are two distinct approaches toward rural development taken by Indian NGOs. The first has as its basis a concentration on the mainstream practice of addressing the symptoms of poverty and underdevelopment by delivering a range of integrated programs and projects, with the involvement of NGOs and grassroots organizations as implementors. According to Farrington and Lewis (1993: 93), there is a consensus among these NGOs and governments which operate under this approach, that "the existing social and economic structures, although inherently important, [are] not directly challenged."

The second, contrasting approach is based on the idea of social action—that effective rural development requires a fundamental transformation of the existing economic and social structures established and maintained by the state. Under this politically charged approach (which, as has been stated, has been officially approved by the GoI), the central role for NGOs consists of mobilizing the poor to make direct challenges and demands on government for equitable service delivery, improved wages and working conditions, and other entitlements (Unia 1991). This focus, which takes a closer look at the underlying causes of poverty and underdevelopment (in contrast with the mainstream program/project approach of addressing the resulting physical manifestations), can be seen in the proliferation of issue-based groups representing such concerns as women's issues, child labor, occupational health hazards, and displacement of populations by large-scale development projects, as well as those advocacy groups that work directly on behalf of the political rights of the rural poor (see Sheth and Sethi 1991). The growing focus on popular participation and social action is, according to Robinson et al. (1993: 93), a major point of difference between NGOs and the Indian government. Although the approaches have fundamental differences, some NGOs believe that the two are not inimical and "that material improvements and empowerment are complementary and can reinforce one another" (Riddell and Robinson 1995: 34). While an NGO which adopts the social action approach to improving rural peoples' conditions may find it difficult to enter into cooperative or collaborative relationships with certain GOs, this does not automatically preclude such possibilities. As mentioned above, one facet of the GoI's official policy is to strengthen the role of NGOs in enabling the rural poor to make demands on local governments, and as one of the case studies in Chapter 5 specifically indicates, the two approaches—mainstream and social action—have been simultaneously employed by the same NGO.

NGOs in Tamil Nadu, where there are approximately 25,000 societies registered with the state government (although perhaps only 5 percent can be identified as rural development NGOs—see Robinson 1991: 25), generally exhibit the attributes of Indian NGOs in general. The majority are small, poorly funded and short staffed, and are generally are not known outside the immediate area in which they operate. Some NGOs operate primarily as contractors to mainstream government-led efforts, while others subscribe to the "social action" approach to development. Many do both. While some look to state and national government agencies for the bulk of their operating expenses, other tenaciously refuse to have anything to do with GOs, usually citing the corruption and hostility of government bureaucrats as their reasons for shunning all connections, even though the scope of their operations is severely limited as a result.

Many other NGOs in Tamil Nadu are much more willing to accept government funding, although an unknown number of NGOs in the state exist largely to enhance their own development, rather than that of the rural poor. Distinguishing the dedicated, competent NGOs from the corrupt is not easy, yet the establishment of meaningful NGO-GO collaborations requires that organizations be identified as suitable stakeholders for such alliances.

Conclusion

In sum, the NGO sector in developing countries is composed of a greatly heterogeneous, largely unassociated amalgam of organizations which share little but their "non-governmental" status and their professed mission to work on behalf of the rural poor. The nature of the roles they play are also subject to a variety of viewpoints, from those who see them as obstructionist (or, at best, a necessary evil), to those who see their significance in society as "Necessary-to-Governance Organizations" (Judge 1995: 178). And, as has been pointed out in this chapter, NGOs are far too diverse a set of actors in the development process to be easily categorized. The ways that individual constituents of such a diversified sector interact with others involved in the conduct of development interventions differ; yet as a whole, NGOs have a significant bearing on the way that development efforts occur in any given society.

Questions regarding the ability of NGOs to expand their current operations and acquire new capabilities notwithstanding, the possibility of GOs and NGOs working together in any sort of interrelated fashion raises a separate issue: whether these organizations can operate as partners in the rural development policy process. The record of NGOs and GOs working together to address problems of rural development is not an extensive one. Relations between the two have been, and in many countries continue to be based more on feelings of mutual antagonism than on trust and cooperation. Recently, though, a growing number of governments have sought the inclusion of their indigenous NGOs in various aspects of their development efforts, and calls for increased cooperation have come from academics and practitioners. In part, these governments have overcome their hesitation to trust certain NGOs. How can the number of such cooperative efforts be increased?

If, as a number of authors have suggested, joint NGO-GO working relationships have the capacity to improve efforts at effective problem-solving, there should be an interest in creating such relationships in a number of countries, under a variety of political, economic and social contexts. In order to do so, the nature of such relationships—what they consist of, and how those involved approach them—must be understood. The purpose of the next chapter is to review these forms of relationships, to identify a further stage of collaborative endeavor—critical collaboration—and to look at the current status of NGO-GO relationships in India.

Chapter 4

Critical Collaboration

> In many respects, by arguing that they need closer relationships with each other, both GOs and NGOs are responding to the recognition of their own limitations. Yet they approach such relationships through perspectives shaped by their histories. GOs still act as if government maintained its central role in administering rural development; NGOs perceive government on the basis of their traditional oppositional relationship with the public sector.
>
> Bebbington and Farrington (1993)

The purpose of this chapter is to present the concept of critical collaboration as an approach to addressing the problems of rural development—not merely as an ideal type, but as an existing form of cooperation between two sectors of society. Collaborative—and especially *critically* collaborative—relationships between NGOs and GOs are not a new phenomena; however, they are not widespread in developing countries. In order to understand their nature, it is important to identify what collaborations consist of, what makes critical collaboration distinct from other relationships, and factors that both facilitate and restrain the creation of these cooperative forms.

In an effort to explain the concept of critical collaboration, the chapter begins with a general discussion of NGO-GO interactions and a specific focus on the views of each actor toward working with the other. This is followed by a review of some specific forms of working relationships that occur between NGOs and GOs. The concept of critical collaboration, and the essential nature of such a relationship which distinguishes it from other forms of cooperative endeavor, are discussed. Finally, a brief review of the history of NGO-GO interaction in India will be presented as a means of understanding the emergence of collaboration in that country.

NGO-GO Interactions

In examining forms of interrelationships between LDC governments and their NGOs, some understanding of the attitudes of the two actors toward each other is a necessary precondition. While the diversity found among development NGOs, as well as the distinctive characteristics and behaviors of national and subnational governments, makes any simple compartmentalizing or typologizing difficult, certain trends and similarities should be recognizable and allow a degree of categorization.

NGO Views

As previously discussed, the attitudes of NGOs and GOs toward one another are, in many cases, marked by a spirit of mutual mistrust and/or hostility. Governments often attempt to maintain control over the development process in the effort to limit perceived threats to their own legitimacy and continuing existence by restricting political and social unrest among their citizenry. At the same time, NGOs often feel that the scope and content of their operations (or, ultimately, their survival) are under constant threat by governments, especially if they are advocates for change to the status quo. In such an atmosphere, what drives NGOs to seek or accept a working relationship with the government?

Small, autonomous, rural development NGOs generally conduct a limited number of activities in a small geographical area, with limited resources (whether financial, material, technical, or administrative) and in relative isolation. As they succeed in their efforts, they look for ways to increase and widen the impact of those successful interventions to include a greater number of people, a process known as "scaling out."[1] As John Clark (1991: 73) writes, "Only a complacent organization would be content to continue with the same job in perpetuity when it is self-evident that its contribution is not more than a drop in the ocean, however excellent a drop it may be."

Given their meager resources, such NGOs find it difficult to expand their operations; to do so, they generally must look to other organizations for assistance. Cooperation with other NGOs or GROs to replicate their programs is one source of such assets; many NGOs also look to their government, or to international donors, to fund their activities. As Farrington and Lewis (1993: 96) write, most Indian NGOs prefer government funding to that of foreign donors for several reasons: first, they do not have to register under the Foreign Contributions Regulation Act, which requires regular auditing and disclosure procedures from the NGO; second, many feel that the "legitimacy derived from government funding" insulates them from harassment by local interests; and, finally, the "strong nationalist tradition" that exists among NGOs "abjures foreign funding on the grounds that it undermines their independence and limits their freedom to determine programme priorities."

Reddy and Rajasekhar (1996) note that NGO-GO linkages are valuable to NGOs for other reasons as well. First, the flow of funding from outside donor agencies to NGOs is not as reliable as it has been in the past. Indian NGOs, many of which have come to depend on outside funding sources (approximately 90 percent of all NGO funding in India is from outside sources [Farrington and Lewis 1993]), face the uncertainty of

---

[1] Julie Fisher (1997: 75) distinguishes between "scaling out," the process by which NGOs widen the impact of their efforts through multiplication of their activities at the local level, and "scaling up," by which NGOs attempt to increase their influence on government policies.

interrupted financial support, especially due to the recently increased attention on development in Africa, Eastern Europe and the states comprising the former Soviet Union. Farrington and Lewis, writing specifically of the Indian experience, note that the choice to work with the government gives NGOs access to a large body of resources. The GoI spends a much greater sum of money on rural development programs than the total amount of NGO receipts from the outside donors. The total amount of NGO funding received from all sources equals only about 10 percent of the GoI's total expenditures (Reddy and Rajasekhar 1996: 10). Cooperative efforts with NGOs could make it more likely that these government resources reach their intended recipients than they would through government programs alone.

Funding is not the only reason for NGOs to cooperate with their governments. It is also advantageous because the state

> control[s] the wide frameworks and policy environment within which people and their organisations have to operate. This not only enables the poor to assert their rights and [helps] them in getting benefits from the government but it also ensures that the grassroots' experiences flow and are integrated into the planning process (Reddy and Rajasekhar 1996: 10).

In addition, NGOs that work in cooperation with their governments, rather than relying on foreign funds, may find their own legitimacy enhanced, "A continuous dependence on external donors may raise questions relating to legitimacy and accountability of NGOs" (Reddy and Rajasekhar 1996: 10). As Clark (1995: 596) points out, NGOs which depend heavily on foreign assistance—especially those NGOs whose activities are politically controversial—may raise suspicion on the part of local elites, government and nationalistic political parties on the grounds that the NGO is being "guided by a foreign hand."

NGOs thus confront a dilemma: while there are real benefits to associating with government agencies, doing so may result in a number of unwanted changes to the NGO's operations.

Association may result in the dilution of the autonomy and flexibility for which NGOs are heralded: participation of outside organizations, especially those with the superior financial, legal and/or political power, may compromise an NGO's ability to design, influence, and monitor the programs in which it participates (Sollis 1991, quoted in Bebbington and Farrington 1993). The possibility of cooptation, and the resulting weakening of NGOs' effectiveness and accountability to their rural constituents, is also a real possibility. In the most extreme examples, an NGO may be decertified or disbanded by its government.

Another, more subtle—yet, according to Farrington and Bebbington (1993), more profound—consequence may also result: that in working with government, an NGO may lose its institutional identity. NGOs which have spent considerable energy and resources in opposing and criticizing government agencies may find that to suddenly be working in a cooperative manner with those same, or other, GOs creates profound tensions regarding the organization's changing purposes and roles.

In situations where the balance of political power is overwhelmingly in favor of government, cooperation is often likely to take the form of cooptation. Many NGOs feel that only in an atmosphere of complete and scrupulous autonomy from government can they serve the interests of their constituents (see Rahman 1984; Fernandez 1987; Paul 1989). NGOs may choose to forego working with government agencies out of the fear of oppression, or because they wish to avoid the hierarchical and bureaucratic structures of government in favor of maintaining their grassroots connections (Fisher 1997: 108).[2]

On the other hand, in forgoing opportunities to work with government, NGOs risk the chance of losing the kinds of

---

[2] Fisher (1997: 109) notes that there is a positive side to an NGO's stance of self-imposed isolation: that strengthening an autonomous civil society may lead to participation and democratization. Such a stance "may have a long-term impact if it leads to alternative approaches to sustainable development and strengthens the autonomy of civil society. Moreover, through their work with grassroots organizations, NGOs can promote political participation and democratization at the local level."

organizational linkages essential to long-term program sustainability,[3] as well as the resources available through official channels (which is especially important to small, poorly funded local organizations). There is evidence to indicate that NGO activities conducted in isolation are frequently unsustainable, especially as funding levels decrease (see Clark 1991).

An NGO which decides to work with a government agency, for whatever reason, must thus make a choice: whether to accept a role within the range of existing policies, or to act as what John Clark (1991) calls "the unreasonable NGO"[4] and attempt to change the status quo, to pursue change in the policies and practices that fail to serve the needs of the poor. If the NGO takes the former course, it will most likely become just another subcontractor engaged in the implementation of a government program. NGOs are often asked to improve aspects of existing programs, but are seldom allowed to design projects themselves or to redefine a program's overall objectives (Clark 1992). Some of the very qualities for which NGOs are lauded go unused.

Clark (1991: 104) discusses the rationale underlying the approach that the unreasonable NGO takes toward government-involved development interventions:

> The unreasonable approach is to seek reforms in policies, attitudes, and practices so as to eradicate the obstacles [to development]. ...[NGOs] are well placed to realize that the self-reliant development they aim to promote will be sustainable only if the policy environment allows it to be, hence

---

[3] L. David Brown (1991: 810) writes that "long-term viability often depends on linkages between projects and other organizations whose cooperation is essential to preserving and expanding their efforts. Some development agencies frequently see other actors (local government officials, village elites, for-profit firms) as a part of the problem. They often protect their fledgling projects from contacts with those agencies, even though long-term sustainability of project gains require their support."

[4] In coining the term, Clark (1991: 104) paraphrases George Bernard Shaw (1990), who wrote that "the reasonable man adapts himself to the world; the unreasonable one persists in trying to adapt the world to himself. Therefore all progress depends on the unreasonable man."

the need to weed out local and national policies and practices which thwart these ambitions.

Over time, many NGOs have adopted values and approaches to development that question and challenge the mainstream paradigm. Clark refers to such NGOs as radical, and writes that these NGOs have come to the realization that development is "fundamentally about taking sides"—that while efforts at improving services and economic opportunities for the poor have had some impact, significant improvement of their lives relies on a policy of large-scale redistribution, not only of incomes and productive assets, but of political power as well, where "Such redistribution is highly political and controversial" (Clark 1991: 37).

Following the unreasonable course—taking sides against the status quo—can be threatening to the well-being of an NGO, especially in situations in which the government is suspicious of, or antagonistic to, the NGO sector as a whole. Yet, as a number of authors have asserted, this may be the most effective way of bringing about authentic development.[5] Taking sides with the poor may also be acceptable to some governments; those which feel confident in working with NGOs, in the spirit of empowering the rural poor, will be more likely to accept NGO critiques as constructive. In many cases, however, such criticism is often seen as an unwelcome challenge to government's legitimate role and is fraught with dangers for an NGO.

As mentioned earlier, the role of the unreasonable NGO is frequently described as an advocate or catalyst on behalf of the rural poor. Some see advocacy as inherently antagonistic toward government; after all, the NGO is consciously attempting to change policies and practices favored by the

---

[5] Clark (1991, 50) writes that "True development is about removing the barriers which restrain people from achieving their full productive capability. This is threatening to those who come to see the erosion of their status, and they are likely to fight back. The only socially neutral course of action is to do nothing. Any intervention alters power balances. NGO's business is, inescapably political. It is about taking sides, siding with the poor against those who comprise the barriers to just development."

government.[6] This role is, for Adil Najam, a natural one, and common to all NGOs; it is an analogue of the government's drive to coopt NGOs, and should be accepted as a normal component of the policy process. He writes (1996) that the very nature of NGO-GO interactions is characterized by a tension arising from each party's need to bring the other closer to its own viewpoint and that the goal of all involved in the policy stream is to move other actors' thinking to reflect more closely their own point of view.[7]

Government Views

How do governments of developing countries view (and, in turn, behave) toward their indigenous NGOs? Such perceptions clearly determine the potential for critical collaboration between the two. Governments generally take one of the several attitudes toward their indigenous NGOs, which are influential in creating the political contexts within which the NGOs are allowed to operate.

[6] As several authors have noted, increasing numbers of indigenous NGOs are taking up the cause of social transformation. As Bhatt (1985: 76) writes, "[m]odern voluntarism strives to change the social, economic and political position of the poor, the deprived, the oppressed, and the weak. In the final analysis ... it aims at the redistribution of power, wealth and status." (see also Riddell and Robinson 1995; Korten and Quizon 1995). Lakshmi Jain (1991, quoted in Heyzer 1995: 206) views this role as crucial; she writes that "When there is no torment [imposed by an NGO on a government], then that NGO can be considered dead. It means that it has no life, no autonomy, no independence. But the very fact that it causes ripples, and that the state is not able to accept it with ease, let alone swallow it, shows that the NGO is really succeeding in...mobilizing the strength of the people."

[7] Najam's understanding of the concept of cooptation is perhaps more encompassing than other authors'; he writes that "[t]here seems a strong normative tendency in the literature to label NGO efforts to use the resources they command (monetary as well as political) to influence government agency decisions as a 'catalytic' role but view government's use of its resources (monetary as well as political) for the same purpose as 'cooptation'." Moreover, many authors, especially those writing from the perspective of the aid-givers, "tend to highlight the dangers of NGO co-optation by governments but have very little to say about the co-optation of local groups by international donor agencies and NGOs, which is not only implicitly condoned but seemingly encouraged" (Najam 1996: 17ff, 16ff).

There is a widespread consensus that NGO-GO relationships are largely determined by one of the several basic attitudes of LDC governments. According to Tandon (1991) there are three broad categories. The first consists of those states characterized by dictatorships, military rule, or authoritarian governments. Most NGOs in these states are viewed by their governments as part of the opposition to the state, and relationships of governments toward their NGOs reflect such an adversarial view (see also Farrington and Bebbington 1993: 49; Tandon 1987).

States with a single ruling party (common in many African countries) comprise a second category. Such states may accept those NGOs engaged in welfare and service roles, especially where the governments have withdrawn from providing those services. In some instances, governments may tolerate NGOs only to the extent that they support their own programs. The state, however, may well be suspicious of any autonomous structures, and growth of the NGO sector in such countries is usually slow, if allowed to occur at all. At best, the state may have an overall policy toward development, but may focus only on one or several issues of their national development while having significant blind spots towards other areas (Tandon 1987).

The third category includes liberal, multi-party democratic states. It is here that the relations between the state and its of NGOs are more complex and difficult to define. In such situations, NGO-GO interrelationships vary; they may be cooperative, hostile, or may change over time. This is a political environment in which both governments and NGOs play mixed roles. Especially in states with complex, multilevel governments, NGO-GO relations have become increasingly multifaceted. Governments may profess concern with the poor while, at the same time, favor certain vested interests over other groups and individuals. NGOs often have antagonistic relationships with certain levels or agencies of government while, at the same time, cultivate and find sympathy, cooperation and assistance from others. GOs also behave in widely different ways with individual NGOs and may exhibit differing attitudes to the same NGO over time (Turner and Hulme 1997:

Ch. 9). It is in this context that cooperative and collaborative relationships are most likely to occur. India, a federal democracy with a diverse variety of state governments, is clearly included this category.

If, as has been suggested in previous chapters, many governments so jealously protect their primacy in the area of development, why should they seek the assistance of their indigenous NGOs in the conduct of development interventions? Why should LDC governments work with organizations that are often seen as potential or actual adversaries? Several reasons exist. Farrington and Bebbington (1993) identify three broad motivations that elicit the widespread interest in NGOs, all based in contemporary critique of the mainstream approach to development. First, governments and donors are looking for more efficient ways of achieving orthodox goals of development. Second, as a result of critique from a number of venues which stress that development should be more participatory, inclusive and diverse, NGOs are seen as a path to possible alternative approaches to development. The third motivation regards a perceived lack of concern with the deterioration of the natural environment: in the past, mainstream development efforts have undervalued the environment and its long-term sustainability. Many environmental advocates see NGOs as more concerned than governments with the environmental issues, including sustainable development and agroecology, and argue in favor of expanding NGOs' roles (see Altieri 1990). All of these reasons are relevant to the Indian situation—the first and second are stated policies of the GoI, and the third a critique of the recent history of that country's development efforts (Khator 1991; Krishna 1996).

In fact, GOs and NGOs do work together in several ways. What forms of interrelationships characterize their working relationships, and what can be accomplished in each?

Forms of NGO-GO Interaction

With the current growth and visibility of the NGO sector in many LDCs, it is difficult, if not impossible, for governments to conduct rural development efforts in complete disdain of their NGOs, just as NGOs do not operate in a vacuum. Riker (1995b) writes that, as the NGO sector grows in size and influence, governments can neither ignore nor coerce their indigenous NGOs without risking considerable political costs, and that governments will have to engage their NGOs, whether in a spirit of toleration and acceptance of the others' independent action, or through involvement of NGOs in their own (governments') development initiatives. Several approaches that LDC governments take toward their NGOs are identified below. These approaches are not mutually exclusive; a GO may use more than one of these tactics with the same NGO or may approach different NGOs in different ways.

According to Julie Fisher (1997), various governments respond to the growth and politicization of their NGOs in several ways. She lists a range of reactions and policies generally taken by LDC governments toward their NGOs. At the most restrictive end of such a continuum are efforts of governments to repress, harass, or otherwise actively limit the activities of their NGOs. A less aggressive approach (although, as mentioned, increasingly difficult to accomplish) is to ignore their NGOs. A third means of government control of NGOs is through cooptation, which takes a number of forms: subtle forms of government bribery; the selective involvement of certain, favored, NGOs to participate in government programs; or, in what is becoming a common strategy, creating their own "nongovernmental" organizations. Finally, many governments take a fourth position toward their NGOs by passively accepting their ongoing activities—tolerating, but without attempting to change or to learn from, NGOs.

At the other end of the continuum lies a fifth approach: interaction (see Figure 4.1). Interaction between GOs and NGOs, which Farrington and Bebbington (1993: 127) characterize as "situations where the actions of one institution are influenced by, dependent on, or oriented toward the actions of another institution," includes a range of interrelationships which may be mutually supportive, adversarial, or something in between. The fact that two agencies are interacting does not imply that the relationship is a positive one, but the fact that some form of contact is taking place may indicate that the activity may be constructive.

Figure 4.1
Levels of NGO-GO Interaction
(adapted from Farrington and Bebbington 1993: 128)

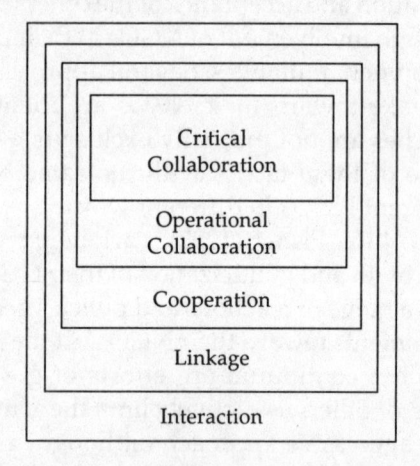

```
┌─────────────────────────────────────┐
│ ┌─────────────────────────────────┐ │
│ │ ┌─────────────────────────────┐ │ │
│ │ │ ┌─────────────────────────┐ │ │ │
│ │ │ │      Critical           │ │ │ │
│ │ │ │    Collaboration        │ │ │ │
│ │ │ └─────────────────────────┘ │ │ │
│ │ │       Operational           │ │ │
│ │ │      Collaboration          │ │ │
│ │ └─────────────────────────────┘ │ │
│ │         Cooperation             │ │
│ └─────────────────────────────────┘ │
│            Linkage                  │
└─────────────────────────────────────┘
             Interaction
```

As a specialized form of interaction, *linkages* between organizations are the most rudimentary form of positive interrelationships. Such relationships include some form of contact, which are either neutral (such as information exchange) or mutually supportive (which may include the exchange of ideas, joint field visits, or discussions), and may be formal or informal arrangements. At the more formal end, such relationships are often characterized by the independence of one partner on another for the successful completion of an agreed-upon activity.

Cooperation is a more formal type of interaction, and distinguished by the use of some kind of mechanism to manage a relationship between the participating bodies. Forms of cooperation vary according to the specific activities at hand. They tend to focus on welfare and service activities, and reflect the government's attitudes and policies toward development—the orientations to development which have proven to be less than adequate in reversing the trend toward increasing poverty and other manifestations of underdevelopment.

Cooperation frequently takes one of two forms, according to Farrington and Lewis (1993: 22). The first, *substitution*, takes place when NGOs implement activities that the government normally performs but for some reason cannot or does not; the GO may lack the resources, expertise, or personnel to operate in certain areas, or the NGO may possess specialized knowledge or access unavailable to the GO. The role of the NGO is that of a contractor implementing a government program, and there is little room for innovation or adaptation of the activity by the NGO.

The second form is *complementarity*, by which both GO and NGO conduct different activities leading to a joint objective. Generally the two make distinct, but coordinated efforts; for instance, the GO might provide hardware and technical inputs, while the NGO conducts social organization and training. In both forms of cooperation, the NGO "will largely continue to be a passive recipient rather than a proactive agent in defining and shaping the course of the development process" (Riker 1995b: 119).

Another, more complex, form of cooperation is *collaboration*. In addition to the cooperative features of the previous activities, there is a greater degree of interdependence between GO and NGO in this form of interaction. According to Farrington and Bebbington (1993: 128), the idea "implies a formalized dependence of one partner on another for at least part of the success of its activities." Samuel Paul (1988, 1989) points to three roles for NGOs engaged in operational collaboration: 1) mobilizing demand for government services; 2) planning and delivering services; and 3) innovating and scaling up NGO initiatives. To this list of what constitutes collaboration, Michael Bratton (1990) adds a distinct dimension: the role of NGO participation in the policy arena, where the NGO may work with a government agency to: a) redirect the allocation of public resources, and/or b) to reorient the focus and/or content of a government program.

Bratton's addition is a significant and an important new dimension in the definition of collaborative activity, for it signifies a challenge to the very core of development orthodoxy: government control of development. By including other actors in the policy process, the authority to shape the direction and content of rural development policy becomes a more democratic and pluralistic process. This is at the core of critical collaboration.

Critical Collaboration

The option for an NGO to become more directly involved in the policy process is central to identifying the concept of critical collaboration and distinguishing this form of collaboration from its operational form. What Barbara Gray (1989) describes as collaboration—and what is here identified as *critical* collaboration, to distinguish it from other operational forms—is "a process through which parties who see different aspects of a problem can constructively explore their differences and search for solutions that go beyond their own limited vision of what is possible." The process of jointly searching for solutions to a problem is an essentially political phenomenon because, by entering the collaboration, each stakeholder is agreeing to share some of the decision-making power in order to come to a solution that is jointly satisfactory. Collaboration

> creates a process by which the stakeholders themselves must wrestle with the question: How can I satisfy my interests in the context of what is in the collective good? Thus, collaboration urges a distribution of power among those whose interests are most keen. ...With a collaborative approach, the final agreement defining the common good is not the proclamation of a ruling elite or the results of political logrolling and majority rule, but rather a consensus agreement among those chiefly involved (Gray 1989: 118).

This factor, of actors' reciprocal and voluntary power sharing, clearly distinguishes collaboration from those activities in which an NGO merely facilitates, implements, replicates, or complements predetermined government programs. Participation in the implementation of a government program does not necessarily give the NGO any decision-making role, although such linkages may serve as precursors or adjuncts to greater cooperation and may lead, as advocacy and other, less friendly stances, to more collaborative relationships.

Gray (1989: 15) compares the nature of collaboration with cooperation and coordination. In comparison to collaboration, the latter two are "static patterns of interorganizational relations"; collaboration is a far more dynamic phenomenon. As Gray points out (1989: 15), collaboration is not a preordained state of organization but an emergent process:

> To presume that the parties in a collaborative effort are already part of an organized relationship underrepresents the developmental character of the process and ignores the delicate prenegotiations that are often necessary to bring stakeholders together initially.

The collaborative relationship is, in and of itself, a more complex and interdependent one than the forms of cooperation earlier discussed, even in the form of operational collaboration, a relationship that may require of each party only the fulfillment of its agreed-upon role. When the component of shared decision-making—including critique and advocacy—is included, the relationship becomes far more tenuous. Collaboration stems from advocacy, whether that advocacy originally manifests itself in an antagonistic stance on the part of the one party challenging the other/s, or in a more moderate form, such as the wish of one partner to enter into negotiations with others over some aspect of a program. It is the agreement of two or more concerned parties to work together despite the differences that exist between them. As such, individuals, groups, and organizations involved in the process must be prepared to adapt to changes in the various circumstances in which they are working, as well to the other stakeholders in the process.

Next, collaboration is based upon a model of shared power: the power, seen as the capability, or capacity, of each partner in the collaboration to make a contribution to the solution of a problem. It is fundamental that each stakeholder in a collaborative relationship is perceived by the other(s) as possessing such a capacity, and that it can be used in conjunction with the capacities of the other collaborating party/ies to bear significantly on the situation at hand. It is evident that all stakeholders see the solution of a particular problem as being in their own fundamental interests (although the individual stakeholders' interests are not necessarily the same) (Gray and Wood 1991: 160–161), and that each anticipates a potential benefit accruing from addressing the problem at hand through mutual action. In addition, each partner must perceive the "interdependence with other groups as necessary for the social problem to be addressed effectively" (Logsdon 1991: 26). Without these two expectations, there would be no incentive for an organization to expend the energy and expense necessary to collaborate with others.

In addition, every stakeholder must understand that it does not have within itself all the capabilities or resources necessary to bring about an effective solution to the problem at hand. If this were not the case, the process of collaboration would be nonsensical: "It is precisely because stakeholders hold countervailing sources of power and their fates are interwoven that collaboration is made possible" (Gray 1989: 112). Power need not be held equally by all participants, but large discrepancies in the degree of power held by each stakeholder will make collaboration more difficult to achieve. In situations in which the power distribution is highly unequal, "the weaker parties must first develop their capacity as stakeholders. They need to establish some form of countervailing power. The extent of this power building will depend on the degree of asymmetry in the relationship" (Gray 1989: 119; see also Nemeth 1970).

As Gray notes, the determination of who is best equipped to participate in a given collaborative relationship is crucial to the process of collaboration, as well as the final outcome. First, a variety of sources of information regarding the problem at hand generally leads to a better, more complete understanding of the issue—a more comprehensive understanding is achieved as the various stakeholders voice their perceptions of the issue and how potential solutions will affect them, and stakeholders "gain a greater appreciation of the patterns of interaction that underlie their actions with respect to one another" (Gray 1989: 64).

The inclusion of those who will be responsible for implementing any decision is also important. Implementors may have crucial knowledge about the feasibility of chosen solutions, and can often identify shortcomings in proposed solutions before the actual implementation phase occurs. Finally, the recipients of any intervention should be included in the process. "Acceptance of any solution is enhanced when those who must abide by it are included in designing the solution" (Gray 1989: 64).

How, then, are certain groups identified as valid stakeholders and potential collaborative partners? What individuals and organizations should be allowed to participate in the processes of decision-making and implementation?

Issues of Legitimacy and Capacity

Gray (1989) discusses two interrelated factors important to the process of choosing among potential partners for collaboration. The first is determining the *legitimacy* of those purporting to have the right to be included in deliberations regarding the issue under consideration: who is entitled to participate? One claim to legitimacy can be made by those who will be directly affected by those decisions made and implemented by other stakeholders. These people often become involved due to their interest in moderating any possible negative outcomes they may experience as a result of the action. Since the majority of problems are multidimensional—their solutions usually affect more than one individual or group, and often in different ways—it is important to gain an understanding of the multiple consequences of proposed actions. Those individuals and groups with the expertise to act on the problem may also be considered legitimate stakeholders.

Regarding this point, the question may be raised about the legitimacy of NGOs themselves as stakeholders. As most indigenous NGOs are not self-help organizations, and are usually urban- rather than village-based (see Chapter 3), they are not subject to many of the impacts of their development decisions to the same degree as their rural constituents are. Nor do they have the ability, in most cases, to authoritatively veto development decisions made by government agencies. However, it would be difficult to expect that most villagers or GROs themselves possess the power to confront, or to work with government at a level approximating that of even the smallest NGOs. It is especially for this reason that NGOs are often chosen to work in the interest of villagers. As mentioned above, the majority of Indian NGOs (as elsewhere) act on behalf of the rural poor, enhancing their potential to engage the more powerful sectors of Indian society. It is clear that this role for NGOs, as a sector, has been accepted (and even advocated) by the Government of India, as indicated in the Seventh and Eighth Five-Year Plans (see Chapter 4). Indeed, "in India, it is now official government policy to encourage [NGOs] to organize the poor in order to make bureaucracies more responsible" (Fisher 1997: 64). Despite the affirmation of the NGO sector's activities by the Central government, though, it is at the local level—of the individual GO, NGO and the particular program—that the question of legitimacy becomes critical. It is at this level that most NGO-GO interactions occur. Not only is it important that NGOs are viewed by GOs as credible actors, they must be deemed equally so by those whose interests they claim to serve (and not seen as merely adjuncts of a corrupt government agency).

As discussed above, the nature of collaboration is inherently political in that it is an exercise in the sharing of power between the organizational stakeholders involved in rural development decision-making. As such, the stakeholders who approach each other in the name of improving rural conditions must possess, or have access to, some degree of power, in the form of the capability, or *capacity*—to a degree that is perceptible to their counterpart—which will justify their claim to be active, accountable participants in a collaborative effort.

As explained in Chapter 3, the government, market, and NGO sectors each possess, through their roles in society, differing forms of power to accomplish a given goal. While the resources accessible to, and the power wielded by, governments in the conduct of development are generally readily apparent, those of the majority of NGOs are often less evident. For the purpose of making themselves credible partners in the policy-making and implementation of rural development, NGOs must be perceived by their potential partners as possessing some tangible attributes—capacities—that are recognizable as being real and useful contributions to solving the problem at hand.

Those capacities necessary for NGOs (as well as GOs) to justify their involvement can take several forms. One of the most commonly acknowledged is that of *technical competence*, the wherewithal to competently accomplish what the organization sets out to do. According to Korten (1986: 13),

> the [NGO] which presumes to help [other] organizations become more effective must be guided by more than good intentions. Having the technical capacity to obtain the respect of those who control the relevant technologies—whether they be doctors, engineers, lawyers, politicians, administrators, or village leaders—is basic. And not being able to buy access to key officials purely on the basis of the financial resources they can offer, they must win access through the perception that they offer a useful technical and political resource.

Another form that is necessary to rural development, which Fisher (1997) refers to as *social and managerial knowledge*, involves an organization's understanding of local management approaches and participatory innovations, as well as social organizing skills. Through field experience in working with the rural poor, NGOs frequently, either unconsciously or as the result of deliberate research, accrue a reliable data source regarding local resources, past and existing problems, and solutions that can be applied to subsequent efforts (Dichter 1986). In addition, the NGO enhances its own understanding of participatory management at the grassroots level.

Another, crucial, area of capacity is the NGO's knowledge of itself and its sense of place in its surrounding political milieu. *Strategic competence*, what Korten (1986: 13) refers to as "a measure of the organization's ability to position its resources to achieve its objectives," is an ongoing, long-term process of self-management, through which an organization consciously creates and strengthens its own position with regard to the political environment in which it operates. Strategic management is primarily directed at relationships between the organization and its external environment. Through repeated efforts at defining and reinforcing a sense of its own purpose, capabilities, and goals, as well as its position relative to the constantly-changing environments within which it works, an organization can position itself to better adapt to changes and crises in its operational milieu, as well as gaining a clear notion of its mission and to better undertake internal change. NGOs that lack a clear and realistic view of their own capabilities and shortcomings often fall prey to becoming involved in activities for which they do not have the necessary capabilities or resources.

The idea that countervailing power is at the basis of collaborative relationships does not automatically infer a struggle for control; collaboration is not necessarily a confrontational process. Like the choice of self-isolation, the decision to enter into a collaborative effort often stems from an NGO's unwillingness to be coopted or controlled by government. Rather than disengaging itself *from* government, collaboration is an effort on the part of an NGO to directly communicate *with* government about what it views as important issues in development. How that communication takes place, however, varies. While collaboration is generally thought of as a positive working relationship, the process may be the outcome of less than amicable beginnings. In a situation where a GO has been unresponsive to public outcry, NGOs may find it necessary to take a confrontational stance to government in voicing their concern about a situation.[8]

The idea that openly adversarial stances taken by NGOs and GOs can evolve into more collaborative ones has been supported by a number of cases in which such a collaboration has taken place. In the Indian State of Karnataka, a consortium of more than hundred NGOs challenged the State's complicity with private industry in the felling of major areas of forest for commercial purposes under the guise of a social forestry program (see Chapter 6). In Sri Lanka, a network of NGOs convinced the government to drop plans for the construction of a coal-fired power station and eventually were invited to participate in the government's National Environmental Council (Clark 1991).[9] In Bangladesh, the government, NGOs and donors have developed a form of working relationship which Sanyal (1991) calls "antagonistic cooperation," a kind of relationship in which the two continue to work with each other in implementation of development programs and projects, despite harboring misgivings and distrust of each other.

It is important to understand that despite all the discussions about government attitudes towards NGOs and the widespread tendency of governments to coopt their NGOs, the choice to work together (aside from some antagonistic forms of advocacy, which—at least at the outset—are not decisions to work *together*) is a conscious one on the part of both sets of actors. Although an NGO may have fewer options than its government counterpart, its decision to participate cannot be coerced (see Najam 1996).

---

[8] Clark (1995: 535) writes that NGO-GO consultation can be a "surprisingly productive process" even when the relationship between the two organizations has been antagonistic and that, conversely, a relationship that has had a history of close cooperation may actually lead the NGO to too readily accept the government's stance and fail to act on behalf of their grassroots constituents.

[9] According to John Clark, the results of the Sri Lankan NGO consortium's (initially) confrontational stance included not only the abandonment of the original plan and the inclusion of NGO representatives on the National Environmental Council, but also the establishment of a process of environmental appraisal for all major projects with possible adverse environmental impacts, and a less confrontational stance between NGOs and the government (Clark 1991).

The tensions and contradictions which appear in the attitudes of GOs and NGOs toward one another in many developing countries—and examples of cooperation amid widespread apprehension—are evident in contemporary Indian development efforts. A brief review of such cooperative efforts follows.

NGO–GO Interaction in India

The status of contemporary Indian NGO-GO relationships results from a centuries-long experience of voluntary efforts in the country, including the role of the colonial government. As Sheth and Sethi (1991) point out, the idea of NGOs as institutions visibly distinct from the government and also the market sectors of society was not widely recognized in India before the existence of a modern Indian state. The creation and recognition of a distinct national entity made it possible to draw distinctions between the actions of the government and those welfare and developmental activities which originated outside the state structure and within society.[10] The early colonial Indian government did, however, recognize the effects of the existence and activities of social reform (and, increasingly, anti-colonial) activities taking place outside its purview. The Registration of Societies Act was passed in 1860 in an effort to regulate and oversee the activities of such groups, especially those involved increasingly in political activities and which made demands on the state to "[enact] new social legislation for implementing reforms.... These... social reform movements transformed the old parallelism between the state and society into a direct interaction between the two" (Sheth and Sethi 1991: 51). The Act is still an active statute used by the GoI in the regulation of NGOs.

The visibility (although not the conduct) of voluntary activities diminished in the first two decades immediately following Independence in 1947. While stating the need to rely heavily on the contributions of a range of voluntary organizations in

---

[10] Although the Gandhian influence led to an earlier consciousness of this role among Indian NGOs (see Sheth and Sethi 1991).

the process of nation building, Nehru's belief was that the state was the necessary and appropriate agent for democratization and social change. While the existence of their work was acknowledged, voluntary organizations were viewed mainly as providers of social welfare and relief services, while rural development was the responsibility of the government.[11]

What emerged from India's early experiment in socialism was a broad network of patron-client relationships sponsored by the Congress system of state control over the economy. Under the control of Nehru's socialist state, many voluntary associations moved closer to official channels of development, increasingly relying on government funding, to the point "that gradually little difference could be seen between their work and the government programmes of social welfare" (Sheth and Sethi 1991: 53).

A growing perception of the ineffectiveness of the government-led nation-building modernization model in the 1960s resulted in a widespread reaction on the part of (mainly middle-class) young people, who moved to India's rural areas and poor urban communities to work as political activists (Kothari 1993). Such political activism on behalf of the poor and marginalized in India was not an innovation of the 1960s. More accurately, it was a resurgence of the anti-colonial activities that flourished in the nineteenth century and the Gandhian self-help *Sarvodaya*[12] movement, which experienced a sharp decline under the hegemonic rule of the Congress Party. What was novel about the new movement was a sense of

---

[11] The GoI, according to Fernandez (1987: 44), has had a long-standing interest in mobilizing people's participation, and has "even considered this dimension essential if poverty were to be removed." What Fernandez characterizes as a "love-hate" relationship between government officials and Indian NGOs stems from a difference of opinion among officials as to whether the role of NGOs in organizing and motivating the people and their institutions (notably panchayats) is an appropriate one.

[12] Sarvodaya, meaning "the welfare (or 'uplift') of all", was initially a Gandhian effort to improve the conditions of the poor and oppressed, and "essential if the people were to be ready for and worthy of [self-rule]" (Rudolph and Rudolph 1987: 132); the concept "emphasized service, nonviolence, and a political and economic decentralization focused on the village" (Hardgrave and Kochanek 1993: 202).

disillusionment and frustration with conventional political institutions, a realization that the promise of Independence and the socialist state to ameliorate the inequities and human suffering was not being realized and that, in fact, the living conditions of the poor had become even worse.

The search for an alternative to what had become a consistently ineffective national development bureaucracy, with its massive, centralized, top-down planning processes, grew. As new social issues—including environment, gender equality, peace and human rights—began to surface in addition to the long-standing concerns, discussions came to be couched, to a growing degree, in terms of politics: if the social and economic inequities that led to rural impoverishment and misery were the result of the political configuration of the country itself, their alleviation could be improved only through substantial change to the structures and institutions currently supporting that system (Sheth and Sethi 1991).

The NGO sector, which had been relatively obscure, slowly grew in stature. Its importance to India's development was officially recognized in the Seventh Five-Year Plan (1985–1990), which "marked a watershed in the way that NGOs were viewed by the government" (Robinson 1991: 32). In the Plan the GoI, acknowledging the limited success of its own efforts at lessening poverty and recognizing the significance of its voluntary organizations, called for a formal inclusion of Indian NGOs in the government's development efforts. For the first time, the GoI called for NGO participation in policy implementation; in addition, the Plan provided one and one-half billion rupees for NGO funding (Robinson 1991: 32). The Eighth Plan took the idea of human development and the involvement of NGOs even further (Dhesi 1996: 205). The Plan called for "creating and strengthening the institutions which facilitate the involvement of [the rural] people" for the purpose of increasing local participation in all aspects of the development process.

While the GoI has been calling for increased NGO activity and popular participation in rural development and increased funding for the same, it has been relying on existing legislative controls as well as newly created ones designed to monitor, and even to restrict, NGO activity. These include the following:

- The requirement that all voluntary agencies with seven or more members register with the government under the Societies Registration Act of 1860. Trusts must do so under the Indian Trusts Act of 1982, and religious organizations must do the same under the Charitable and Religious Act, 1920.
- The passage of the Foreign Contributions Regulatory Act (FCRA) of 1976, in response to allegations that subversive activities were being funded through outside sources (and strengthened in response to ongoing political unrest in the Punjab). Through the Act, all NGOs wishing to receive funds from abroad must register with the Home Minister, to submit audited accounts twice yearly, and to detail all foreign donor contributions. It is seen by some as a pretext for governments to investigate NGOs that are supposedly affiliated with political opposition groups (Robinson 1991).
- The merging, in September of 1986, of two existing organizations—People's Action for Development Initiatives (PADI) and The Council for the Advancement of Rural Technology (CART)—to form a semi-autonomous organization, the Council for People's Action and Rural Technology (CAPART). According to Mark Robinson (1991: 32), while the body was created with "the twin aims of promoting NGO involvement in rural development and promoting new technological innovations through NGOs," it plays a politically powerful and controversial role. CAPART administers funding from a number of Central government programs for use by NGOs[13] including several environmental and natural-

[13] The majority of NGO funding comes from foreign donors; only about 10 percent of their funding comes from GoI sources. Most of the money channeled through CAPART is from ministerial budgets, although a small proportion comes from bilateral aid sources. However, the use of either domestic or foreign funds by NGOs is not without some controls, detailed guidelines and limitations.

resource programs.[14] Moreover, CAPART has succeeded in coopting much of the spirit of the cooperative movement in India and despite the inclusion of NGO representatives on its board, operates largely independent of their input (Verma and Menon 1993: 331).

The GoI has attempted, over a number of years, to institutionalize NGO accountability through the creation of a code of conduct for those NGOs working with governments. In 1986, the GoI attempted to create a National Council of Voluntary Action; this met with strong protest on the part of many Indian NGOs (Tandon 1993). Subsequently CAPART, in a conciliatory tone, suggested in a preparatory document for the Seventh Five-Year Plan that "[NGOs] may evolve their own Code of Conduct [which] should provide for adherence to a rational, secular and scientific approach to the issues of people's concerns on the one hand and subservience to the democratic ethos on the other" (CAPART n.d.: 7).

These proposals have been the source of an ongoing debate (dating from the mid-1980s) regarding the appropriateness of such attempts. Rajesh Tandon (1993: 200), a critic of NGO regulation, views the creation of a code as a means to stifle the autonomy of NGOs: "the State and its machinery [have] attempted to regulate that type of voluntary action which has refused to become subservient to the development policies of the State." Opposition to such government regulation is based on several assertions: that the intent of regulation is an "attempt ... to undermine the voice, the space and the commitment of independent and autonomous voluntary action" (Tandon 1993: 201); that the spirit of voluntary action conducted by NGOs extends beyond the narrow criteria of government regulatory control based on funding—" voluntarism as more than a kind of constructive bureaucratic or developmental work" (Baxi 1986: 28); and that

---

[14] These "include social forestry, ecological development, ... [and] the provision of safe drinking water ..." (Robinson 1991: 32).

the [asymmetries] of power are so sinister that any process of extending state regulation to the voluntary sector will further arrest genuine processes of empowering the impoverished, ... the quintessence of democratic social action by voluntary agencies (Baxi 1986: 32).

Others see the need to regulate NGOs as vital to the spirit of voluntary action in India. One of the most outspoken proponents of a code of conduct, Bunker Roy, has been in the forefront of the effort to regulate NGOs, primarily due to his perception of threats to the voluntary sector arising from widespread corruption. According to Roy, the result of massive flows of funding from outside, combined with a lack of NGO accountability, has led to the destruction of the "spirit of volunteerism" in India (Roy 1996; see also Roy 1993, 1995; Shourie 1995).

A number of scholars recognize additional positive aspects to the institutionalization of NGO-GO relations. Farrington and Bebbington (1993: 52–3) point out that beyond the issue of political control, the issue of NGO registration with their government has practical considerations. Coordination between the two could prevent the redundancies and conflicts that occur when a number of NGOs are implementing activities "which overlap and conflict both among themselves and with those of government." Edwards and Hulme (1996b: 8) also see positive aspects to the creation of institution-wide accountability measures for NGOs. They write that the only way the NGOs can avoid corruption by other organizations "is to develop systems for performance monitoring, accountability, and strategic planning that assure that a line remains between transparent compromise and blind co-optation" (see also Eade 1993: 161).

In sum, the overall policy of the Indian government presents certain constraints to the establishment of close working NGO-GO relationships. At the same time, the individual states pose a very different set of limitations.

The Indian States

Within India, there is a divergence of NGO-GO relationships among the states. Although there is an overall government policy of favoring closer working relationships and the strengthening of NGO activities at the national level, there is considerable variation in the way state and local governments behave toward working with NGOs operating within their borders. This is in large part due to the variation in state governments, many of which are ruled by regional parties with distinct ideologies and policies regarding the roles of government and other actors in development.

In West Bengal, for instance, NGOs have been frustrated in their efforts due to the limits placed on their activities by the long-ruling Communist Party in that state, which sees NGO activities to empower the poor as a threat to its own political authority (Fisher 1997: 57–8) and to its socialist ideals. As a result, the state government has undertaken to decentralize power, but only in line with its own ideology. One such action has been to institute a number of laws designed to restrict the activities of NGOs (Rajasekhar 1997). State rule by a government hostile to the GoI has had national repercussions. For instance, antagonism between the Central government and West Bengal's leftist parties has led to the GoI's hesitancy "in providing funds and resources for NGOs in West Bengal because of the fear that they will be utilised in building up the cadre of Leftists or that the credibility for rural development efforts will be attributed to the [Communists]" (Kalimullah 1989: 97).

A short-lived effort was made by the Government of Rajasthan in the early 1980s to actively involve NGOs in the development process, not only to monitor and provide feedback on government programs but also to participate in "highlight[ing] controversial issues of conflict ... which could be attended to quickly, thus re-establishing the credibility of the delivery system of the Government" (PRADAN 1996b: 25). A consultative group was created for the purpose and existed for several years. Despite the enthusiasm of high-level officials, the program foundered due to the failure of the government to convince the lower ranks of the state bureaucracy regarding the merits of the effort. Despite the fact, that today there is no clearly defined policy on its relations with NGOs, the state maintains an ongoing project with a large indigenous NGO to study the nature of existing NGO-GO relationships in that state, as well as future possibilities (PRADAN 1996b).

In Karnataka, the state government (GoK) has shown a decided interest in developing working linkages with NGOs. In addition to making internal, political, and administrative changes intended to promote decentralization of its rural development, the GoK has created both state- and district-level consultative committees. These committees include NGO personnel as members and on the recommendations of a working group on NGOs; a number of joint sub-committees have also been created to encourage further practical interaction between NGOs and the government (see Bhat and Satish 1993). This favorable atmosphere has resulted in the creation of a growing number of state- and district-level NGO networks, some of which have as major goal participation in the state planning process and playing a cooperative role with GOs (see Rajasekhar 1997).

In Tamil Nadu, NGO-GO relations are embedded in a longstanding context of mistrust. Moen (1991: 83–4) points out that as throughout India and many other countries, government agencies in Tamil Nadu attempt to control those NGOs they view as threatening to their own interests: from more "friendly" forms, such as subtle bribery intended to reward compliance by favored NGOs, to more dire adversarial actions such as harassing NGOs workers, denouncing their activities to local police and Central government agencies, and threats to or actual physical abuse of NGO workers.

Collaborative ventures are growing in that state, yet, as is the case elsewhere, a large number are the result of the interest of individual local bureaucrats in the establishment of such relationships (who, in turn, must obtain a clearance from the appropriate state-level elected representative) (Rajivan 1997). One feature of Tamil Nadu's government is important in this regard. Political parties in power in the state have, for several decades, been decidedly pro-agrarian, and state policies have encouraged NGOs there to experiment on different approaches to rural development to a greater extent than has been the case in some other states (although higher-level state policies, in Tamil Nadu as elsewhere, do not necessarily filter down to the local level). The state government has also worked with its own NGOs in several initiatives to improve social security in that state, including school meal programs, innovative pension plans, and improvements in access to health care (see Drèze and Sen 1995).

Conclusion

In this chapter, the nature of several forms of interactions existing between NGOs and GOs in India and other developing countries were reviewed. One form, termed "critical collaboration," in which both parties join together in a conscious effort to work as partners in formulating and/or implementing rural development interventions, was suggested as a emerging and promising form of cooperative association between government agencies and indigenous NGOs (though the extent to which critical collaboration actually occurs is largely undocumented and has not achieved any established status).

In the effort to identify and understand the existence and occurrence of critical collaboration between Indian NGOs and GOs, a study of a number of actual cases that have taken place was conducted in the state of Tamil Nadu during 1996 and 1997. The next chapter is a discussion of the characteristics of each of the NGOs that entered into collaborations with several GOs, and a narrative of the events leading to the relationships.

Chapter 5

Case Studies

> Both the state and the public have central roles to play in economic development. While different schools of thought tend to place different emphasis on their respective roles, it is hard to avoid seeing them as thoroughly interdependent. Just as the nature of state activities depends a great deal on public demands and pressures, the actions of the public—both collaborative and adversarial—are all the more effective when the state plays its part in helping to empower the citizens by guaranteeing basic democratic freedoms, ensuring widespread literacy, protecting the rights of disadvantaged groups, and providing some security against extreme destitution. There is, in this sense, a deeply complementary relationship between state action and public action.
>
> Jean Drèze and Amartya Sen (1995)

In the previous chapter the nature of NGO-GO interactions, and some of the social and political contexts that shape the ways that such interactions take place, were discussed. In particular, some of those features especially applicable to India were highlighted. In addition, the concept of critical collaboration as a distinct form of complex, deliberate, and interdependent action was presented and described. This chapter is an account of several examples of relationships created between NGOs and GOs in the Indian state of Tamil Nadu. Through a review of the individual NGOs, GOs, and the events that occurred between them, these case studies intend to identify the important features of the relationships and whether each constitutes a critical collaboration.

The research, conducted in the state of Tamil Nadu during 1996 and 1997, focuses on five NGOs and the corresponding GOs involved in joint efforts to address problems of the rural environment. Both personal interviews and secondary sources of information were gathered about each NGO, including age, personnel, organizational structure, sources of funding, past and current activities, and a specific program involving interaction between the NGO and one or more GOs. In each case an ongoing environmental/natural resource issue in which both the NGO and one or more agency of the State of Tamil Nadu and/or the Government of India were currently involved was identified. In each case the potential for the creation of, if not the actual existence of, a relationship of critical collaboration was evident. The studies were designed to explore both the basic attributes of the NGOs and GOs necessary for the establishment of critically collaborative relationships and the conditions shaping the actions of the actors. The relevant features of the five cases, each distinct in several important attributes, are presented here in a narrative form. Chapter 6 analyzes and compares these findings, and examines the extent to which the interactions between the NGOs and the GOs can be considered as relationships of critical collaboration.

NGOs involved with environmental and natural resource issues in Tamil Nadu include a wide range of organizations with a variety of attributes. These organizations vary in regard to a number of characteristics: in size, from those with offices and operations in several states to the office with only one or two paid employees; in the resources available for their use (both tangible and intangible); in the number and variety of issues with which they are involved; and in their orientation toward working with the government (or eschewing such connections)—and, as a result, their approach to government officials in general. The NGO sector in Tamil Nadu, as in India and other countries, is far from homogeneous.

Despite their differences, all the NGOs operate within largely common, physical, political and social contexts. Tamil Nadu is a semi-arid state in south India that presents a number of environmental and natural resource problems for its rural population. Availability of agricultural water is a real difficulty for the majority of farmers. For instance, according to Meinzen-Dick (1984), the majority of the state receives an annual average rainfall of less than 950 mm (38 inches), almost all of which falls during the monsoon months of September to December. As in much of the country, forest cover in Tamil Nadu has been severely depleted, and deforestation continues largely unabated despite a number of national and international policies and efforts to halt the high rate of destruction. The climate is largely undifferentiated (except for the higher elevations of the Western Ghats) and varies little throughout the year, with the exception of the monsoon season. Also, like the rest of India, approximately three-quarters of the population lives in rural areas, where the majority are landless peasant farmers. Social stratification is highly pronounced and rigidly maintained.

All the NGOs studied here are involved to a large extent, if not solely, with environmental and natural resource issues in Tamil Nadu. Each has shown, through its efforts at development, the ability to engage individuals and agencies of the government in the conduct of environment-oriented activities (although in different ways).

The NGOs vary significantly in several characteristics (see Table 5.1), a number of which have been identified as important to the creation of collaborative relationships. The attributes of each organization, including its size, age, level of funding, range of activities, and orientation toward development have been noted at the beginning of each case. In the presentation of the case studies, a review of the program studied and a brief narrative of how the program as well as the interaction between the NGO, GO and any outside organization involved is conducted has been provided. Each case study has a distinct record of how its interactions with government agencies have taken place. Analyses of the cases, and how they represent the creation of critically collaborative relationships, appear in Chapter 6.

Table 5.1
NGO-GO Case Studies

NGO	Year of Creation	NGO Age*	Annual Budget (Rs)**	Principal GO/s Involved	Type of Project
PRADAN	1992	9	50,000,000	DRDA/PWD	Irrigation Tank Restoration
AWS	1986	9	2,080,000	DRDA	Watershed Rehabilitation
PHCC	1990	5	4,900,000	TNFD	Afforestation
SPEECH	1992	5	5,000,000	TNFD	PRA Education
SHIELD	1994	2	300,000	NAEB	Afforestation

* At the time of the inception of the program studied.
** 1997 figures

Professional Assistance for Development Action (PRADAN)

The largest and most diversified of the organizations studied, in terms of the area of operations as well as the variety of development programs in which it is involved, is PRADAN. PRADAN is one of a relatively few, large, and multifaceted NGOs in India which conduct a variety of operations in a number of states. Now based in Delhi,[1] PRADAN has an annual operating budget in 1996 of more than Rs (Indian Rupees) 5 crore (50 million rupees, or approximately US$ 1.4 million),[2] a significant portion of which was provided by international donors, including the Ford Foundation and the European Union. The organization has (unlike the majority of Indian NGOs) a wealth of resources, including its own office buildings in several states,

[1] At the time of the study, the Madurai office served as the national office of PRADAN, originally housed in Delhi. Since that time the Delhi office has resumed the role of headquarters, while the director of the Madurai office has "spun off" that branch to continue the south Indian operations of PRADAN under the auspices of a new organization, Development of Humane Action (DHAN).
[2] Source: PRADAN (1995). Conversions of Rs to US $ reflect the exchange rate at the time of the fieldwork, approximately Rs 35-US $ 1.

extensive libraries, a number of vehicles, and computer facilities. PRADAN is well known throughout the state of Tamil Nadu's NGO community, and has working relationships with a number of NGOs and other groups, as well as with local, state, and national GOs.

PRADAN conducts operations in nine Indian states with a full-time paid staff of more than seventy-five people working with 31,000 families in more than a thousand villages. PRADAN's scope of rural development activities includes programs in natural resource management (including land and water rehabilitation and management, agroforestry, wasteland development, and small-scale irrigation), income-producing microenterprises (such as tasar silk production, poultry rearing, mushroom cultivation, and animal carcass processing), and creating and maintaining self-help groups for savings, microcredit, and self-employment. More recently, PRADAN became involved in a Government of Tamil Nadu (GoTN) program to investigate and address the high levels of female infanticide in a number of areas in rural Tamil Nadu.

PRADAN was created in 1983 through the efforts of several university-educated rural development practitioners and the Delhi office of the Ford Foundation, an international NGO [INGO]. Their motivation to create the organization was based on two convictions: first, that the non-governmental sector in India had "an important catalytic role in complementing government efforts at rural poverty alleviation" (in sharp contrast to the more confrontational social action groups that proliferated in the late 1970s). Second, they felt that the effectiveness of many NGOs could be enhanced through "professional inputs in a wide variety of technical and managerial fields" (PRADAN 1997: I–4). PRADAN's approach was to address both concerns in a less ideologically driven, and more pragmatic, atmosphere than either the Gandhian spiritualism or the radicalism of the social action groups. PRADAN was founded to serve as a catalyst, rather than an antagonist, toward Indian government at all levels.

PRADAN's initial approach as a catalyst was: to identify young development professionals with relevant managerial, organizational, and technical skills; to train them in the practical aspects of rural development; and with personal and professional support, to place the individuals in other NGOs with which they would work to improve the agencies' capacities to conduct their work. The purpose was to integrate the individuals as members of the NGOs, and they were to work directly with the organizations while retaining access to PRADAN for organizational and technical support. PRADAN calls this process "professionalization from within"; the core idea is that through experience, the individuals would over time gain the skills and experience necessary to be able to work without the formal support of PRADAN, maintaining only an "alumnus" status with the latter. At the same time, the involved NGOs could continue to look to PRADAN for technical and managerial assistance as they expanded their own capabilities.

The founders surmised that this second agenda, enhancing NGOs' capacities, was a natural and logical concomitant of the first. The lessons learned through experience, however, indicated that this was not necessarily the case. Young and inexperienced professionals, they found, lacked the skills necessary for managing organizational development; PRADAN itself was "built largely with people with little or no grassroots experience, little organizational experience and a lot of formal knowledge" (PRADAN 1996a: III–8). At the same time, attempts to place the professionals with outside organizations were less than completely successful. PRADAN professionals themselves often felt that they were serving two masters, were resented as outsiders by the NGO staff, and often did not experience the kind of personal and professional development that PRADAN had intended to impart.

In response to this perceived shortcoming, PRADAN augmented its original mission of institutional development. In addition to the original strategy of assisting medium- and large-sized organizations, they added a focus on smaller, local voluntary agencies; professionals were to act as outside assistants, rather than as resident members of the organizations. The focus of the professionals changed as well, with a greater concentration on the technical aspects of specific programs in small geographic areas than previously dealt with. The approach was perceived to be preferable to the original one in several ways, especially on the part of local and community-based groups which placed a higher priority on PRADAN's ability to provide technical and financial skills than on the NGO's need for organizational growth and change.

PRADAN's original emphasis on issues of organizational improvement continues, even in these small groups. In addition to the provision of technical and administrative expertise to individual NGOs, PRADAN also serves as a forum for local NGOs, assisting in the development of organizational attributes within individual NGOs which will improve their cooperation and beneficial working relationships with those at the grassroots.

The directors of PRADAN realized from the organization's outset that the role of NGOs in influencing the policy process would be crucial. An early progress report (December 1984) states that

> If their innovations [are] to make a dent on the condition of the poor, voluntary agencies need to devise mechanisms by which they can influence policy. PRADAN professionals therefore, spend some time in translating their project level work into implications at the policy plane (PRADAN 1996a: VII-5).

Through both research and field experiences, PRADAN has maintained a continuous involvement in the study of NGO participation in the policy-making process (see PRADAN 1986, 1988, 1990, 1991 for titles regarding their policy studies). A study of PRADAN's participation in a recently initiated statewide program focused on rebuilding a traditional method of crop irrigation and watershed rehabilitation provides an insight into the NGO's relationships with the government.

Case Study: Tank Irrigation in Tamil Nadu

One of PRADAN's more recent and extensive efforts to participate in the government policy processes has been in the area of natural resource conservation: the reconstruction and long-term management of a number of traditional irrigation tanks located throughout rural Tamil Nadu.

Given the seasonal and erratic nature of rainfall throughout the region, the use of various methods of acquiring and storing sufficient irrigation water is necessary in nearly all areas of south India. Irrigation tanks are one such time-tested technology, used throughout the region for storing rainfall and runoff for a number of uses: crop irrigation, groundwater recharge, domestic use, cattle watering, pisciculture, and other uses. Tanks consist essentially of an empoundment pond created by the construction of a man-made bund (embankment) on a hillside or slope, with channels to transport water to the users. The length and height of the bund, and thus the size of the tank, vary according to the slope of the land and the anticipated rainfall/runoff. Irrigation tanks have been remarkably efficient for the storage of runoff and were historically constructed throughout south India, as well as in Sri Lanka and other countries. Tank irrigation currently accounts for more than 30 percent of all irrigated cropland in the state of Tamil Nadu. In the states of Andhra Pradesh, Tamil Nadu and Karnataka it is estimated that today there are about 140,000 tanks, in various states of repair. There exist approximately 39,000 irrigation tanks of various size in Tamil Nadu alone, of which more than 20,000 are relatively small, serving command areas (ayacuts) of less than 40 or fewer hectares (ha) (Saktivadivel *et al.* 1982: 2).

Despite the their long-standing existence, most irrigation tanks in Tamil Nadu have fallen into neglect and disrepair. As in the other states, large-scale technologically-intensive forms of irrigation (dams and canals) as well as the pumping of groundwater resources from bore wells have taken precedence over less extensive, "minor" forms (tanks and farm ponds), and the bulk of government funding for irrigation purposes has gone to such large projects. Ascher and Healy (1990) point to a number of political and institutional reasons for this bias toward large-scale projects. They state that one of the most difficult problems in many natural resource issues, and especially in the area of irrigation, is that of equitable distribution of benefits—both regarding the even-handed distribution of water for those in the command area (the area immediately served by the irrigation project) and those outside the command area. In addition, there are technical difficulties; tanks require constant maintenance and upkeep to keep in proper repair. The result of these forces is that minor irrigation is underutilized in proportion to its potential (see Sanpathy 1984: Ch. 4).

The disregard of tank maintenance is not only a recent problem; an 1868 report of the British Chief Engineer of the Mysore State wrote that

> One thing to my mind is perfectly clear, namely that no department or collection of departments can even hope to keep going the ordinary maintenance of the ... tanks in this province. The work is at once too large in the aggregate, too small in detail—as well in fact [to] attempt to keep every hut in the country in repair (Vani 1992: 99).

In the pre-colonial times, irrigation tanks were maintained by the individual farmers and village organizations. Historical epigraphs indicate that farmers received grants or rent and tax remissions from local rulers for their participation in tank maintenance and repair (Ambler 1994). The British, however, treated tanks primarily as a source of revenue, rather than as a long-term supply of a valuable renewable resource. They concentrated the expenditure of funds for tank maintenance only on those which returned a given amount of revenue to the State. Others, not as remunerative, were considered a financial burden and not worth the cost of repair, and were thus allowed to languish. Though the British eventually recognized the need for these resources, and attempted to recreate local traditional institutions—"tank panchayats"—for small tank maintenance, but it was an exercise that proved futile.

For a number of reasons, the disrepair of the majority of small tanks and the loss of irrigation water to many of the marginal farmers of Tamil Nadu persists. To a large extent, jurisdiction is a primary cause. The local institutions necessary to the ongoing maintenance of small tanks do not exist (due to the inability of the governments to provide the necessary human and other resources for the work). Moreover, control of the tanks remains largely with the local and state governments. Despite their attempts to assign the responsibility of tank maintenance to local farmers, most agencies are unwilling to relinquish control over the larger tanks. In Tamil Nadu, the responsibility for tanks which serve less than 40 ha of land is, by law, shared by local panchayats and state Water Resources Organizations (WROs), which are a division of the state's Public Works Department (PWD). However, many panchayats in the state exist in name only and lack the ability to raise the necessary funds for tank management. Control of local irrigation resources remains effectively with local elites and state agencies. Finally, government control of tanks and related components is fragmented among a number of agencies and levels of government (although local governments are responsible for many of the tank structures, the surface water itself is under state control). Any changes to existing practices must be approved by a number of (often competing) agencies unwilling to relinquish control over their area of jurisdiction.

In 1992, PRADAN formally agreed to work with Anna University of Madras (AU) on an ongoing research program investigating distribution problems of water usage. They approached the District Rural Development Agency (DRDA), which was working on the AU program, to request that a portion of the funding be set aside for a research study. PRADAN chose twelve sites to conduct a pilot program, focusing on the specific problems of the smaller tanks and marginal farmers (which had been abandoned by the DRDA and the state).[3]

Following the study, PRADAN requested that they continue their participation with the university and several state NGOs

---

[3] PRADAN officials felt that, because the repair and maintenance of small irrigation tanks is a common property resources issue, any related program activities (including funding) legitimately require participation of the State (Vasimalai 1997).

in negotiating the tank program. The discussions, which took place over a period of several years, focused primarily on how to effectively devolve power to local farmers to develop plans and processes for rehabilitating, maintaining, and managing damaged tanks for villagers' use in a way that would coincide with local needs and practices. PRADAN's approach was to create local Water Users Associations (WUAs) which would have immediate control over maintenance of the tanks and distribution of irrigation water. Their expectation was that local GROs could more effectively manage the day-to-day operations of the tanks and the distribution of the water than a district or state agency.

After months of often-contentious deliberation, agreement was reached with the DRDA; however, additional discussions were required before the PWD agreed to the proposal, and the program could begin (Saltyagopal 1997). According to PRADAN additional issues, including encroachment by local elites, project funding, red tape, and the time frame for project activities, were major obstacles.

The first problem was of encroachment by local elites. It involved portions of the tank beds (waterspreads) that emerge as the water recedes, and land adjacent to the tanks, which in many cases are owned by the government, being frequently damaged by unauthorized and illegal agricultural use. This happens usually due to the activities of local elites or others having access to local government officials, who allow such illegal activities to occur. Also, managers of one tank in a connected series—a tank *cascade*—may withhold water from tanks downstream for their own, unauthorized, uses. Convincing authorities to prosecute encroachers or prevent such activities is often difficult: those who challenge these practices often face retribution from the encroachers or those who hold local power.

The scheduling of project funding was another contentious issue. The usual method of funding public works consists of a schedule of payments to the contractors after the completion of a given percentage of construction. The standard schedule, in this case, was unsatisfactory to the WUA members. In it the workers' pay was not immediately available (a common problem, often forcing would-be workers to take jobs elsewhere), and the method of payment (and the hands through which the money passes) almost invariably resulted in a portion of the construction funds being used to pay off corrupt officials. In this case, PRADAN' size was an asset; using some of its own reserves, the NGO was able to pay village workers while negotiating changes with the DRDA on the schedule and the methods of payment.

A pervasive issue confronting all bureaucratic procedures is the process of submitting proposals and receiving approval from the myriad of involved agencies, which often leads to delays, misunderstandings, and contradictions, and which, in turn, results in bottlenecks or interruptions in implementation. Especially at the local level, agency officials and operatives are hesitant to take any initiative that may stray from the strict state or central policies, and which may threaten their position. In addition, according to one PRADAN official, the corruption of middle-level officials frequently results in the failure to pass on existing rules or changes in policy from higher levels to the local offices (PRADAN 1997).

Government program and project implementation is usually planned in a rigid time frame of several months or a few years. This often coincides with the schedule of elections—politicians showcase the visible benefits of their sponsored programs at the election time (see Ascher and Healy 1990, especially Ch. 7). Many NGOs, on the other hand, realize that the time-frame for their own work, especially in creating and maintaining institutions for long-term project maintenance, can be an extended process whose results are not visible for some time. Reconciling differences between GO and NGO views is usually a long-term process in itself (Shanmugam 1996). All these problems are exacerbated by the temporary nature of District officials' tenures and replacement; they are regularly re-posted every 2 or 3 years to prevent their cooptation by the populations they served (see Chapter 2).

PRADAN has been successful, to a degree, in addressing some of these problems. They were able to create WUAs to manage the small tanks, and persuaded the PWD to revise the schedule of payments to the tank workers. According to one

PRADAN official, these changes occurred because of: 1) PRADAN's past record of successful collaboration with GOs; 2) the commitment of the NGO to long-term involvement in programs, one not limited to project funding cycles; and 3) its willingness to negotiate with relevant government agencies, and sensitivity to GO concerns and difficulties (PRADAN 1997).

The claims regarding PRADAN's success at engaging in collaborative relationships will be evaluated in the analysis of the cases in Chapter 6.

Auroville

Auroville is an anomaly in the midst of south India. It was created in 1968 as an experiment in what was intended to be an intentional "international cultural township" in the environs of Pondicherry, a former French colonial enclave on the southeastern coast of India. According to its founders, Auroville would be a "universal town where men and women of all countries are able to live in peace and progressive harmony above all creeds, all politics and all nationalities. The purpose of Auroville is to realize human unity" (Auroville 1996).

At first glance, Auroville as a whole, as well as its constituent working units, cannot be readily considered an NGO as the others in this study are, in the sense that the NGOs work with a local village or grassroots constituency in the pursuit of efforts to accomplish rural development problems (see Chapter 3). According to one Auroville document,

> Auroville is not a village development society, but an experiment for a large group of volunteers interested in tackling the problems besetting humanity .... In this context, the villagers present a significant challenge but not the whole focus of the Auroville endeavour (Auroville 1998).

However, as will be discussed, the deliberate effort on the part of the Auroville Water Service (AWS) to work with local villages in the role of a legitimate NGO has had repercussions for its ability to work with local GOs as a collaborative partner.

Auroville was created with a great deal of enthusiastic support from both the GoI and UNESCO (its inauguration was timed deliberately to mark UNESCO's twentieth anniversary), and received funding from both in its early years. The connection between the township and UNESCO at the time of Auroville's creation was such that Robert Minor (1999: 102) refers to Auroville as a "project" of UNESCO.

Auroville's overarching purpose, at the time of its inception, was not one of service to the villages with which it is interspersed, but to be an international township and experiment in intercultural understanding and cooperation (although, in its 1984 resolution, UNESCO includes in its activities both restoring the region's ecological balance, and conducting experiments in alternative energy sources and economic development—see UNESCO 1984). Auroville was created with the idea of fulfilling a number of idealistic principles, yet such practicalities as its ownership and how it was to operate as a model of "international unity" were not specified at the time. Arguments over its ownership continued for almost two decades after its creation, until settled by the adoption of the Auroville Foundation Act in 1988 (Government of India 1988). There is, however, a recognition that in addition to their ideals, practical considerations regarding Auroville's existence in the midst of rural India are necessary and that "ways and means of including the material and cultural needs of the villages in the Auroville development planning need to be worked out" (APCG 1996: 17).

The visitor to Auroville will immediately see that Auroville is distinct from the neighboring villages of Tamil Nadu. The land, which was a treeless wasteland, is now lush with native species of trees and other flora and fauna. Widely dispersed buildings reflect European architecture more than local Indian designs. At the visitors' center, one is more likely to find cappuccino, brioche, and lasagna than *lassi*, *chapati*, or *masala dosa*.

The majority of the inhabitants of Auroville are distinct as well. Most are not Indian nationals; of the approximately two thousand residents, it is estimated that more than two-thirds hold foreign passports. Many members of the township—"Aurovilleans"—do not view their work as comparable to that of conventional rural development NGOs. As one long-term member noted, "We really don't work *for* the villagers; we do pay them to help us, and we tell them what we want them to do" (David Nagel 1997). Yet, in the legal sense, Auroville is an NGO: as the director of the AWS pointed out, the Auroville Foundation was founded in 1988 as such, and is registered with the GoI under the Foreign Contributions (Regulation) Act as an NGO, in large part to legally receive the substantial portion of its operating revenues from foreign sources (Gablier 1997).

Over the years of its existence, Aurovilleans have increasingly realized that participatory and cooperative activity with local villagers has become necessary to the operation of the township. As Auroville has grown, its efforts to assist local populations have expanded as well.

As the poverty of the land and the population were clearly interlinked, it was possible for Auroville to make multiple efforts in many areas, each helping the other, despite lack of formal coordination. The good results, such as they are, depend on the people's effort. The people were very eager to work and, for instance, projects to plant trees, bund and fence the land, provided employment and cash to the people as well as directly working with the environmental problems (APCG 1996: 4).

The geographic location of the "universal township" has itself created an unusual situation regarding Auroville's relationship with the governments of both Tamil Nadu and Pondicherry. Although small portions of Auroville are located within the Union Territory of Pondicherry, the township is almost completely surrounded by the state of Tamil Nadu (Auroville is located directly inland from India's East coast; in fact, several settlements of the township are located directly on the shore of the Bay of Bengal). Thus, Auroville, its operations, residents, and political status all present an anomaly in the context of Indian development.

Auroville conducts much of its everyday commerce with the government of Pondicherry, and some of its administrative functions are directly tied to the Territory. For instance, all vehicles are registered in Pondicherry, and Auroville is represented on some of the territory's advisory bodies and working groups (see Giordano et al., in Farrington and Lewis 1993: 147). The township is, at the same time, legally registered as an NGO with the state of Tamil Nadu and many of the Central government-sponsored programs that are applicable to the township are administered through the GoTN. Such dual jurisdictions can be confusing and, at times, problematic for the daily operations of the township.

Auroville contains a number of diverse research and commercial operations, including: environmental regeneration, organic farming, non-traditional education, renewable energy research, new building technologies, and a range of small-scale industries and handicraft productions. The operating units are, to a large extent, autonomous in their day-to-day operations; there is little formal coordination of their functions.

In the three decades since its inception, Auroville has earned a reputation for its numerous contributions to the Indian environment. In addition to reclaiming and reforesting much of the local landscape, the Auroville Greenwork Resource Centre has catalogued and created a seed bank of indigenous species (the *Index Seminum*); it also distributes plants, seedlings and technical assistance throughout India and abroad. Auroville has become a major source of environmental information and innovation in India, in part due to its widespread network of scientific and technological knowledge. Primarily because of the international nature of the township, and its numerous activities with Indian and international bodies, Auroville has become a major, well-established center of environmental expertise in India.

Case Study: The Auroville Water Service and the Local Watershed

The Auroville Water Service (AWS) is one of the constituent working units of the township, created in 1986 in response to Auroville's growing need for water resources. The AWS reflects the orientation of Auroville toward its surroundings in general. As are other constituent units of Auroville, the AWS is a loosely organized working group, consisting of a staff of thirty eight, with thirteen administrative, executive, and engineering personnel and an operational staff of twenty five. Operating revenue for 1997 was approximately Rs 660,000 (US$ 19,000).

Concerned primarily with the development of the township itself, for a number of years the Water Service made no concrete efforts to include the active support and participation of inhabitants of the surrounding villages in its plans for the future of the township. Nor, according to the director and founder, did they make a conscious effort to address needs beyond Auroville before the inception of the current program (Gablier 1997). The basic purpose of the AWS has been to ensure that sufficient water resources continue to be available for the future population growth of Auroville. It was only with the realization of the scope of the environmental threat to the area's ecosystem, and the size and complexity of the required intervention on its part, that the AWS looked beyond Auroville's internal concerns to search for solutions to a growing problem.

As a part of Auroville's planned future growth, the question of sufficient water to support the anticipated population of the township arose in the late 1980s. The deteriorating status of the area's water resources has become a critical issue, threatening the area's ability to sustain human as well as animal and plant life in a coastal area of southeast India. Once a barren wasteland, the phenomenal success of Auroville to reclaim the area and make it productive agricultural land has led to overuse of limited natural resources. The challenge facing AWS is a complicated one.

According to the director of AWS (Gablier 1997, 1998, 1999), water reserves in the area have increasingly been depleted in recent decades. The area was largely wasteland prior to the creation of the township. The successful efforts on the part of the Aurovilleans over the past three decades to rehabilitate the land have encouraged intensive farming practices in the area by a growing local population of villagers as well as Aurovilleans. The growth of agriculture has led to large increases in the use of local groundwater resources. Wealthier farmers have taken advantage of government-subsidized electricity by installing water wells and continuously pumping groundwater, both for their own use and for sale to poorer farmers, with the result that the area's water supplies have been dangerously depleted. The almost constant drawing of groundwater has depleted three levels of aquifers, including the deepest, containing "ancient water" (accumulated over millennia and which require long time periods to replace), with predictable results. In addition to the problem of resource depletion, salinization and waterlogging of the soils in the area (which are the common results of over-irrigation) have increased. Another danger has appeared: the encroachment of salt water from the nearby Bay of Bengal. Aquifers are being recharged with salt water, and the saline content of an extensive local wetland, which is a major resting and feeding point on the migration route for many species of birds, has increased. Under these circumstances the future of the wetland, and the ability of the local environment to sustain agriculture and, ultimately, local habitation, are in question.

In 1995, AWS was approached by an agency of the Canadian government, which offered funding for the Service to conduct an overall survey of the water resources in the area and to suggest possible methods to bring current water use in line with its availability. The director of the AWS prepared a proposal for a large-scale engineering project to increase the capacity of local tanks. The original proposal was rejected because it did not take into account the fundamental causes of the problem or the need to modify local communities' water use practices.

Having had no training or experience in the social aspects of rural development, the AWS director consulted another organization in order to modify the proposal to address the human needs of the area; it was evident to him that the AWS could not do so alone (Gablier 1997). He met with PRADAN in early 1996 to discuss the social and political aspects of the situation, and possible approaches to the mitigation and solution of the growing crisis and its causation. After several days of discussion, AWS and PRADAN drew up a "General Strategy" for a program that would cause minimal adverse social impacts to the area while providing a solution to the overuse of water resources in the area. The strategy include the construction and rehabilitation of twelve irrigation tanks, areawide afforestation projects, a general restoration of the local environment (including the re-introduction of endangered plant and animal species), and the creation of new social institutions for the management of these innovations. The last aspect was almost certainly the most difficult; it involved making fundamental changes in the political and social relations in the area, including the transfer of some control of the resources to a group of previously powerless people who had not learnt the skills necessary to effectively maintain and use that power.

AWS realized that the program—which has been named "Harvest"—would be further complicated by the local political situation. Making the necessary environmental changes would require ongoing engagement with a number of government agencies, each having jurisdiction (and competing to retain their control) over some portion of the planned program activities. This undoubtedly necessitated a certain measure of negotiation and collaboration, along with a way to convince GOs that an alternative to the current situation was in their interests.

Recognizing that the dual requirements of creating cooperative (let alone collaborative) relationships with the relevant, sometimes intransigent GOs on one hand, and with comparatively powerless villagers, on the other, was beyond their expertise, AWS asked PRADAN to act as a sister organization.

It was expected to train the AWS members as an organization which could then act as an intermediary while, at the same time, initiating the processes of building local grassroots institutions and mechanisms capable of maintaining mutual cooperation between the state, AWS and the villages. During this period, the Madras Institute for Development Studies (MIDS), a private think tank in Chennai (the state capital), also requested to become involved in the program for research purposes.

Following the creation of the General Strategy for the watershed's rehabilitation, AWS conducted several preliminary activities with the assistance of PRADAN. First, they visited the involved villages on a number of occasions, in order to: 1) identify possible area leaders who could act as AWS contacts and initiators within the villages; 2) create understanding, enthusiasm and support for the program among villagers; and 3) solicit local information, ideas and concerns. Second, AWS began a concentrated study of the environmental and social conditions in the area in order to tailor components within the overall program to fit more closely the individual needs of each community, and to coordinate relations between villages associated with tank cascades. Finally, AWS initiated a series of discussions which involved state and central GOs in an attempt to foster inter-agency cooperation and coordination on the program. In January 1997 they participated in a "National Seminar on Farmers' Participation in Tank Rehabilitation and Management" sponsored by the GoTN and the European Communities, a sponsor of the AWS program (IMTI 1997). The project is still in the preliminary stages, and relatively little construction work had taken place at the time of the interviews.

Another important facet of the relationship with local governments is the role of the Central government in local decision-making. As explained in Chapter 2, the federal system of Indian government, and its presence in local government in the form of the District Collector and other IAS officials working at the local level, is an important feature of rural development. One of the central partners with which the AWS has created a collaborative relationship is the District Rural Development Agency (DRDA) (see Chapter 2), an autonomous agency in the GoI's Department of Rural Development. The DRDA is, according to the Ministry of Rural Areas and Employment, "the [central government's] implementing agency for all the rural development programmes at the district level" (DRDA 1999). The Director of the AWS indicated that the DRDA has been one of the Central government's primary contacts in the program and that the two organizations have, over the past few years, succeeded in building a collaborative relationship based on reciprocal needs:

> [Our] partnership came out of mutually built confidence, and also a mutual interest. They want results, we provide results, they have the money to spend in tanks and we need money to work in tanks. ...This mutual interest is fundamental to build a partnership (Gablier 1999).

As mentioned, the original proposal for the watershed program was rejected by both State and international funding agencies, and had to be redesigned to better fit the needs of the local populations. AWS turned to an experienced NGO—PRADAN—for consultation and training, and the creation of an approach which would include a holistic view of the communities, resources, and aim at minimization of the adverse impact of the program on local villages (Gablier 1997).

The following cases involve NGOs whose sizes, budgets, and operations are considerably smaller than either PRADAN or AWS. They also lack the wide range of contacts and technical capacities of the larger organizations. How these NGOs have been successful in engaging GOs in productive relationships, despite their lesser endowments, provides some insights into how critical collaboration may occur among those NGOs with significantly limited political power, staff and financial resources.

Palni Hills Conservation Council

The Palni Hills Conservation Council (PHCC) was founded in 1985 as a small *ad hoc* group of concerned citizens—including botanists, environmentalists, foresters and businessmen—living in the environs of Kodaikanal, originally a British hill station[4] in the hills of western Tamil Nadu. Over the intervening years PHCC has grown in size and in scope; today there are sixty three full-time and approximately fifty part-time paid employees, and a varying number of volunteers and technical experts. The NGO is involved in a range of diverse environment-related activities, including the planting and growing of tree seedlings, investigations of water pollution in the area, apiculture, biodynamic agriculture, and wetland preservation. PHCC's operating budget for 1997 was Rs 4.9 million (US$ 140 thousand).

The organization was created by a few individuals who shared a general concern over the environmental degradation that was occurring in the Kodaikanal area, including the impacts of rapid urbanization in the resort and its surroundings. PHCC's first activities focused primarily on local informal education programs, promoting environmental awareness and protection among the populations of the area surrounding Kodaikanal. The members soon realized that efforts at preserving the natural environment in the area required addressing a wide range of issues, many of which they felt "had to be talked about with the local government" (Jeyakaran 1997). Their first experience in engaging the government on environment issues was an effort to prevent the planned uprooting of an ancient tree in Kodaikanal. As they identified more issues regarding the area's environment (to which they felt that the local government had not fulfilled their responsibilities to the public at large), PHCC took an active role in confronting the local and state governments for their inaction in implementing existing environmental laws and regulations.

[4] Hill stations were enclaves in the cooler, higher elevation areas of India created by the British to provide an escape for the colonial rulers from the torrid summer climate of India. Today most serve as tourist destinations.

In 1987, PHCC sought action under the Environment (Protection) Act (1986) to prevent the construction of a complex of vacation homes, on the grounds that the discharge of the resulting untreated sewage into a local marsh that supplied water to the Kodaikanal lake would severely damage the already fragile local ecosystem. PHCC also determined that there were insufficient water resources to support the development, and that the planned boring of water wells would further upset the fragile ecological balance. After receiving no reply from the state government—which a *Times of India* article claimed was in "active connivance" with the developers in destroying the local environment (Bhushan 1998)—PHCC appealed to the Union Ministry of Environment and Forests (MoEF) to force a halt to further construction. An acrimonious exchange ensued between Centre and the state officials, which finally resulted in the cessation of further construction, although occupation of the housing built before the filing of the petition by PHCC was allowed (*The Hindu* 1998).

In 1994, PHCC took a further step toward compelling the local government to fulfill its legal obligations. They brought suit against the city of Kodaikanal and the District Collector for their lack of performance on several counts, notably the city's failure to prevent (and, in the opinion of some, actively authorizing) the construction of a hotel in violation of the city's building code (*The Hindu* 1994). The code restricted construction to buildings of no more than two stories in height. The developers of the hotel, who were constructing a seven-story building, insisted that five of the stories were constructed below grade and were, in fact a multi-level basement. In a hotly contested and widely publicized campaign, PHCC was ultimately successful in convincing the Union government to force Kodaikanal officials to order the demolition of the hotel.

Despite these examples, not all of PHCC's relations with local and state governments have been confrontational; there have been a number of projects on which the NGO and GOs have worked together constructively, although often separately. For instance, PHCC is working with the Kerala Forest Department to develop an elephant corridor linking the Palni Hills with a national park in that state, and has participated in a series of joint wildlife censuses at several national parks as well as in the Palni Hills. More recently, PHCC has succeeded in entering into partnerships with a number of local GOs. According to several officials, this has been the result of the NGO's history of community involvement and their willingness to work cooperatively with local authorities even while, taking individual GOs to task as the occasion arises.[5]

In their activities to protect the natural environment of the Palni Hills, the PHCC has moved beyond the immediate environs of Kodaikanal to include restoration of the ecosystem in the surrounding hills themselves. One recent activity of PHCC, which might be categorized a critically collaborative working relationship has been the Kadavakurichi Interface Forestry Management Programme, an effort to protect the area's resource endowment for the long-term needs of the rural population.

Case Study: The Kadavakurichi Hill Interface Forestry Management Programme

In 1990, the state coordinator for the Swedish International Development Agency (SIDA) approached PHCC with a request that the latter participate in a program that was, at the time, being implemented by the Tamil Nadu Forest Department (TNFD) in the Kadavakurichi Hills, located near Kodaikanal in Tamil Nadu's Western Ghats. PHCC had been working on its own initiative in areas adjacent to the FD-owned reserve forest—planting seedlings, rehabilitating wastelands, and encouraging the participation of local farmers in afforestation activities—with some funding from the GoI, notably the National Wasteland Development Board (NWDB). SIDA, which had been funding the TNFD's Social Forestry Program and was aware of PHCC's work in the area, suggested that the NGO work in concert with the Forest Department, with the

---

[5] Simultaneous activity on the part of an NGO to confront one or more government agencies, while at the same time working cooperatively with others, corresponds to what Julie Fisher (1997) refers to as "acupuncture."

hope that a joint effort could improve the results of the FD's efforts. The PHCC officials responded by stating that they conceived of two ways that the NGO might work with the Forest Department: either to "follow the standard government menu of activities," which would place PHCC in the role of contractor to the FD's project; or (an alternative which PHCC advocated) to look for ways to improve the FD's project by determining the best course of action. PHCC suggested that they undertake a study of the situation and, subsequently, prepare their own proposal for presentation to the TNFD (Jeyakaran 1997).

PHCC determined that the FD's sole effort at afforestation of the Kadavakurichi Hill area to date was to plant a limited number of seedlings within the reserve forest (owned and strictly regulated by the FD), which were quickly destroyed due to the browsing of new shoots by grazing animals. PHCC officials recognized, in part from their own experiences at land management, that more effort was required for successful afforestation and rehabilitation of the area, and that any solution would have to be workable in the context of both the social, and natural environments of the area. They conducted a house-to-house survey of the villages in the area that formed an "interface" between the reserve forest and agricultural land, in an effort to determine alternative solutions for the area's problems. The survey, which used Participatory Rural Appraisal (PRA) techniques, included not only a review of current land use and vegetation cover, but also the local population's agricultural practices: who was using the forest resources, for what purposes, and what people perceived as their own needs. The study, presented to SIDA and the TNFD, recommended a number of new activities designed to rehabilitate the land directly adjacent to the reserve forests so as to provide resources for the use of local villagers without depleting resources within the reserve forests. Their decision to concentrate on rehabilitating the interface area was made, in part, on the basis of the legal status of the reserve forests, which were being illegally encroached by the villagers.

Responses to the results of the study varied by level of government. At the local level, the results were dismissed summarily. Local officials viewed PHCC as intruders into their domain and legitimate jurisdiction, and were uninterested in what they saw as a challenge to their authority. At the district level, there was a sense of curiosity among a number of officials. PHCC felt that, while there was an interest in the study, most officials were unwilling (or felt unable) to deviate from their standard practices, and would continue to rely solely on their own personnel and regulations. It was only at the state level of the Forest Department that serious consideration was given to the results of the study: a number of top-level officials, although not all, were willing to discuss the possibility of changing their policies regarding restoration in the area, including the application of untried innovations.

As the result of several years of negotiation with PHCC, in June of 1995 the District Forest Officer of the TNFD signed a Management Agreement (Tamil Nadu Forest Department 1995) with PHCC. Several of the provisions of the document indicate an uncharacteristic departure from the Department's previous unwillingness toward working with the public, including agreements on:

- decentralizing the responsibility for the program's management—"to be brought under the sole and exclusive control of the [District] Interface Forestry Division" (p. 2);
- sharing decision-making with PHCC as well as the local Village Executive committees (p. 4); and
- submitting disagreements involving "any major disagreement in the preparation of the JFM (Joint Forestry Management) plan [to outside arbiters] for consultation and decision ..." (p. 6).

In addition, the FD agreed to participate in rehabilitating the interface forest areas as well their own reserve forests, and to share the funding of the rehabilitation.

The most important role that PHCC sees for itself in this project is that of a catalyst: creating and strengthening village organizations and grassroots planning committees that will prepare the villagers themselves to participate with FD personnel in implementing the Joint Forestry Management program. As a central component of the Kadavakurichi project, PHCC has established community centers and educational programs (both traditional and non-traditional), farmer training programs, and planning groups in each of the involved villages. As a result of the NGO's educational activities, people who have been traditionally wary about having any contact with the FD personnel (whom they see as the policemen) began to participate in the government programs, although many were still somewhat hesitant about doing so (Jeyakaran 1997).

The Kadavakurichi project, which PHCC began in four villages in 1994, has since been expanded to more than twenty. According to the Director of PHCC, however, the Forest Department has withdrawn some of its participation in the program: "the 'joint' aspect of joint forest management is being abandoned by the FD" (Kannan 1998).

Society for People's Education and Economic Change

The Society for People's Education and Economic Change (SPEECH) is representative of the majority of NGOs in India—it is small, operating almost exclusively in two blocks, or "taluks,"[6] of the state with a total of twenty five full- and part-time operatives. SPEECH, which has a small office in the city of Madurai for administrative purposes, bases its operations in the small village of Tiruchuli, located 50 km south of Madurai, in an arid region of southern Tamil Nadu. The NGO was founded in 1987 in response to the perceived need to promote the self-development of the rural poor in an area suffering some of the most dire conditions in the state: the highest infant mortality rate in Tamil Nadu, excessive exploitation by politicians (even in comparison to other areas), high rates of illiteracy, and prolonged periods of drought (the area is so dry that many of the inhabitants cannot produce any crops, and

[6] The taluk is the smallest unit of government in India larger than the village; there are six taluks in the Virudunagar district of Tamil Nadu.

many families' children—who work in the local matchstick and fireworks factories—are their sole breadwinners).

SPEECH's stated mission is that of "combating rural poverty and socio-political imbalances to improve mass education among the area's rural poor" (Erskine 1999). The founders saw the need to raise local recognition of the underlying causes of the problems of deprivation in the area, and to assist villagers in acquiring the skills to improve their living conditions through their own action. To accomplish this SPEECH introduced a number of community activities: organizing community members for self-help efforts, providing access to preventative health care provision, teaching techniques for local natural resource management (specifically land and water use), and promoting and developing sustainable organic farming practices in the area.

According to its own brochure, SPEECH's approach to accomplishing these ends

> is highly class based rather than caste based. It ... emphasizes the fact that liberation of human potentials cannot be confined to a "bottom-up" oneway process, but initiate a two-way process of transformation at the top to accommodate the changes being generated at the bottom (SPEECH Statement of Purpose–n.p.).

This is a conscious realization that interaction between villagers and local government agencies and bureaucrats is necessary, although local initiatives are generally unlikely to succeed unless the need for change is accepted by the government, as well as by the larger land owners and local elites. From the outset, the officers of SPEECH found that their ability to effectively influence local GOs was limited, largely due to their small size and dependence on the state and national governments for funding. In turn, according to the director, their size and dependence on governments also makes them vulnerable to manipulation by corrupt officials.

Limits to their ability to influence GOs became evident when, in 1989, SPEECH attempted—unsuccessfully—to convince the local Public Works Department (PWD), a state agency largely responsible for the regulation and verification of infrastructure and construction projects (see Chaturvedi 1988) to allow increased local control of a watershed project. They found that their participation could progress no further than that of a contractor and implementor of pre-designed government projects.

However, SPEECH has, in recent years, been able to enhance its ability to work with government agencies despite its limitations. This influence has resulted not from its participation in the implementation of rural development projects or from exhortations to local GOs, but because SPEECH is skilled in a set of techniques that are increasingly seen as an important new way of eliciting local participation in the planning and policy making of rural development: Participatory Rural Assessment (PRA).

Case Study: Participatory Rural Appraisal

PRA has emerged in the past few years as an innovative approach to collecting information regarding the rural poor. Instead of the mainstream approach to gathering data, there is a reversal: the poor themselves are encouraged to take the lead in the process of information gathering, teaching outsiders about their lives, livelihoods, and other information crucial to the design of development interventions. PRA is based on the idea that local people can, and should, conduct the appraisal and analysis of their own lives (see Chambers 1994).

In 1989 SPEECH first learned of Participatory Rural Appraisal (PRA), and what could be accomplished through its use. In that year, the founder of SPEECH, John Devavaram, attended a training session held by a Karnataka-based NGO. He gradually became convinced that PRA could be a valuable set of techniques for use by NGOs and GOs for including community participation in development efforts, through the use of innovative methods for learning information regarding local conditions. He returned to SPEECH and convinced skeptical operatives to attend a one-month training seminar in

Kerala taught by Robert Chambers, a veteran scholar of rural development and one of the creators of the PRA approach. Soon after their training, SPEECH began to train the staffs of other local NGOs and donor agencies in the uses of the techniques.

In 1992 Devavaram was approached by a GoI official who was on assignment as a liaison officer between the GoI and the Department of International Development and Cooperation, Denmark (DANIDA) for the state Forest Department's DANIDA-sponsored social forestry program. DANIDA, like many other international funding agencies, increasingly advocates the use of participatory appraisal techniques in the programs they fund, and advocated the FD's use of PRA in the planning of the program. Since Devavaram was an advisor for the DANIDA-sponsored program (as well as a member of the board of SPEECH), he suggested that SPEECH be asked to provide training. The NGO was contracted to conduct a series of training sessions. SPEECH's position, that of being the sole organization in the region with the expertise to train the FD personnel in the techniques, was a fortuitous one.

At the outset of the FD's training, there was again a great deal of skepticism among the local-level officials about the merits of using such unfamiliar, untested techniques. The training was initiated only at the insistence of the Division-level Forest Officer, an IAS official, who was familiar with the techniques and impressed with their results. As a result of a two-day set of sessions focusing on the practical uses and applications, many of the attendees were convinced that PRA had the potential to be a time- and cost-effective means of obtaining useful information. Forest Department officials who attended the workshops were especially impressed with the quality and amount of information they were able to elicit from the villagers in such a brief, concentrated period of time. According to one official, the FD agents also perceived that villagers, normally fearful of government agents, became much more willing to be active participants when they viewed their role as that of teachers to the bureaucrats, rather than as passive recipients of official edicts (Rengasamy 1997). The reversal of roles, which places villagers as teachers of the outsiders, is a crucial feature of PRA.

The teaching of PRA, says an advisor to the NGO, is context-dependent. SPEECH teaches the techniques at three levels. At the highest levels, he explained, the theoretical aspects of PRA are taught to policy-makers and administrators; at the grass-roots level, the practical techniques are stressed, for use by field workers and district-level officials; and, at the intermediary level a mixture of theory and techniques are taught. The use of PRA in rural development has caught on in many government circles in the state; "even the [District] Collectors are supporting the use of PRA" (Rengasamy 1997).

SPEECH has emerged as a major source of PRA training in Tamil Nadu, and has created a new organization, the Tamil Nadu Association for PRA, solely for that purpose. In 1995, they were asked by the state PWD to take part in a tank systems study "examining hydrological and socio-economic inter-relationships between tanks within a single watershed" in a southern area of the state, because, according to the PWD, they were one of only a few organizations in the state with the capability of conducting PRA (Chandrasekaran 1997: 2–5). SPEECH has been sought by other GOs in the state to teach PRA methods: both the Agricultural Engineering Department and the Revenue Department have requested PRA training. The organization has also conducted workshops at both the national and international levels to teach PRA techniques.

Sempatti Hill Initiators for Ecological Development

The smallest and youngest of the NGOs chosen for this study is the Sempatti Hill Initiators for Ecological Development (SHIELD). SHIELD concentrates solely on environmental issues in a small geographic area; the organization has few paid staff (two full-time and one part-time), and a budget of less than Rs 3 lakh (US$ 8600) in 1996.

The NGO, with a one-room office in the village of Palamedu, located approximately twenty four km north of Madurai, was created in 1992 as a part of PRADAN's ongoing program of fostering the creation of small, local NGOs to address local problems. The intent of the PRADAN officials has been to create small organizations to act as facilitators in forming small functional village groups that can address local environmental problems—in this case, to counter the effects of deforestation in the immediate area of Palamedu. According to their 1995–96 Annual Report, "[t]he main focus of SHIELD is creating awareness among the people on environmental issues and promoting such activities as afforestation, soil conservation measures, developing orchard and agroforestry models" (SHIELD 1996). In addition to this, SHIELD lists as one of their main objectives "promoting viable linkages between Government Department[s] and people." One way of fostering village participation is by including members of each of the villages in which they are operating in their governing board. Another is to conduct education programs in local primary schools, in which university volunteers assist in teaching.

The scope of SHIELD's operations is small, and intentionally so: it was felt that a local organization would work best if its size were carefully controlled, so that its activities and effectiveness would not be overwhelmed by uncontrolled growth. Unlike the majority of small NGOs that focus on a limited geographic area, the directors of SHIELD have made a conscious effort to understand the implications of their limitations—in terms of size, age, capabilities, and access to resources. In order to emphasize the advantages that reflect these characteristics, they have chosen to place limits on the scope of their operations. According to the director, SHIELD will restrict its activities to no more than twenty five villages. Rather than increase their size in the future in order to work with an expanded clientele, they will create new NGOs to handle increases in the number of needy clients. At the time of its creation, the organization worked in only four villages (with populations averaging less than hundred) in a single block of the Madurai district. By 1996, their operations had been expanded to fifteen villages.

Bureaucratically, SHIELD is top-heavy, with a twenty-member volunteer executive committee (seven of whom are farmers and landless, unskilled laborers, known as "coolies"). The day-to-day operations, however, are managed by two full-time and one part-time employee; training is conducted by these people, with the addition of outside volunteers and other NGO personnel, brought in on an *ad hoc* basis to share their expertise. At the same time, the SHIELD board holds monthly meetings to assess the progress of their efforts and to plan upcoming activities; the board includes two representatives of each of the villages in which they are working. The NGO's operational structure is referred to as a "member control organisation" in accordance with its stated commitment to be accountable to the needs of its rural constituents.

Case Study: Afforestation in the Sempatti Hills

As discussed earlier, the GoI, in addition to the powers available through several provisions in the country's constitution, has created a rural development bureaucracy that both duplicates and parallels the responsibilities for various aspects of rural development under state jurisdiction. The experience of SHIELD—their failed attempt to conduct a series of afforestation projects with the state's Forest Department, and their subsequent interaction with an agency of the GoI—illustrates this aspect of India's federal system.

In 1992, their first year of operation, SHIELD requested funding from the Tamil Nadu Forest Department (TNFD) for the creation of a joint forest management project in an area adjacent to an State reserve forest near the village of Palamedu. At the time, SHIELD had begun working on a program for the Animal Husbandry Department to increase fodder production on about thirty hectares of wasteland.

The TNFD denied the NGO's proposal, despite SHIELD's willingness to take the responsibility for all program administration and implementation, and in contradiction to an order from the State Conservator of Forests to comply with the request. SHIELD turned instead to a Central government agency within the Ministry of Environment and Forests (MoEF), the

National Afforestation and Eco-Development Board (NAEB), which has a program for providing grants-in aid to NGOs and schools wishing to conduct activities "directly or indirectly connected with development of wastelands" (NAEB 1996: 1). Funded activities include the planting of seedlings, training and extension programs, soil and moisture conservation works, afforestation, and grass and fodder development.

The NAEB program has a number of grant requirements: only non-profit entities are eligible for funding; local participation (including such local institutions as village panchayats) in both planning and implementation is mandatory; and there are rigid provisos that 1) the sponsoring organization have the capacity to successfully implement the project, and; 2) that

> the established NGOs take on the role of facilitators, innovators and/or motivators rather than the implementors of projects of NAEB assistance. Thus NGOs are expected to motivate people to form into groups at the grass-roots level [and] help them to formulate technically sound & viable waste lands development proposals ... (NAEB 1996: 2).

At the same time, the NAEB outline is relatively flexible in allowing individual recipients to design many of the aspects of their projects, including how the money is to be spent, the methods of training farmers and other participants, the technical details of the project, and when and how it will conduct periodic reviews of the project's progress with the villagers.[7]

In 1993, SHIELD initiated a program with NAEB funding, planting about thirty thousand tree seedlings on hundred acres of privately owned land, to provide a buffer zone to the reserve forest. At the conclusion of the first year of NAEB funding, SHIELD's program was evaluated by a private consulting firm, which transmitted their findings and suggestions to NAEB. Since that first year of the program, NAEB has incrementally

---

[7] NAEB has a regional office in Bangalore, but monitors the progress of its grantees largely through the use of contracted consulting firms which conduct field reviews of the NGOs, their past operations and plans for the projects for which they are requesting funding. The reviewers subsequently file written reports and recommendations with the Delhi office.

increased the funding level to SHIELD; the NGO has subsequently expanded the project size, from four to ten villages.

Conclusion

Chapter 4 explained the nature of NGO-GO collaboration, and suggested that such relationships had at their basis a component of shared power—that, in order for a government agency to accept the prospect of voluntarily sharing a portion of its power over the control of the development process with another actor, the latter would have to possess some countervailing power—in the form of capability, or recognizable attribute(s)—that would make the particular NGO a credible partner in the estimation of a government agency.

In this chapter specific cases of rural development, in which efforts to create critically collaborative relationships between NGOs and GOs were undertaken, are presented. Whether or not such relationships have occurred in the cases will be investigated in the next chapter. The factors identified as important to the creation of the NGO-GO relationships, how each affected the creation of that relationship, and the degree to which the relationship can be identified as a critical collaboration, will be discussed as they apply to each of the cases.

Figure 5.1: Study Area

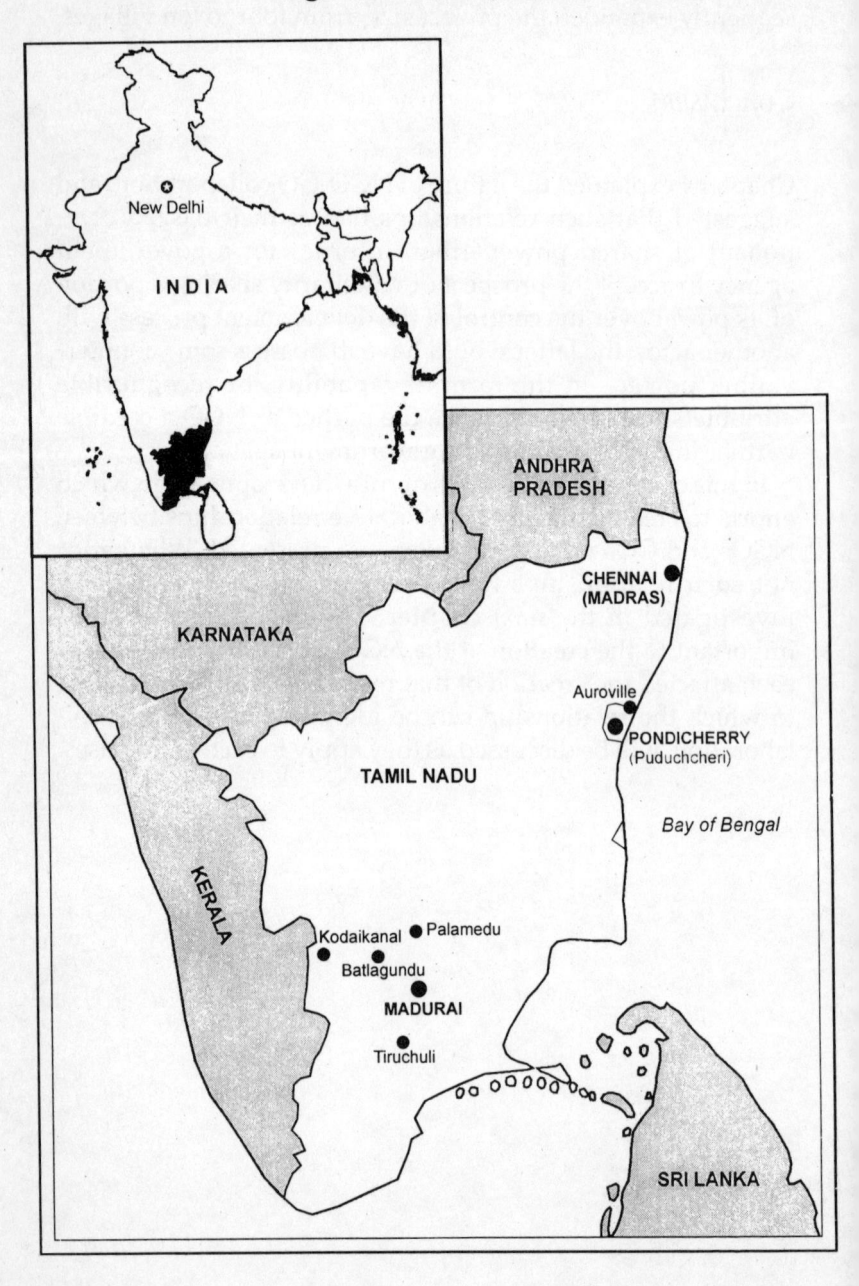

Chapter 6

Collaborative Relationships, Policy-making, and Program Design

> The most important relationship for the NGO sector is with the government. Since both government and the NGOs represent a diverse and confused universe, there is no fixity to their relationship—which is marked both by admiration as also suspicion and hostility.
>
> D.L. Sheth and Harsh Sethi (1991)

Chapter 5 described the cases of five NGOs that have worked with the various levels of government in India in their attempts to halt or reverse the deterioration of the natural environment in Tamil Nadu. Although each case indicates distinct differences between the NGOs, the programs and/or the manner in which the relationship between the NGOs and GOs was initiated, there appear to be similar outcomes: an existing or the possibility of a relationship of critical collaboration. The purpose of this chapter is to identify the factors in each case that are seen as important to the creation of those relationships, and to answer the central question of the study: under what circumstances do NGOs and GOs work together as interactive and interdependent entities?

As explained earlier, a fundamental feature of critical collaboration is the component of shared power, seen as capabilities. A collaborative relationship, involving two or more stakeholders addressing a given problem, is based on a set of mutual perceptions: that each is a legitimate actor; that each is capable of contributing to the solution of that problem; that it is to the greater advantage of all parties to work in partnership than to work separately; and that the stakeholders have agreed to share the authority to both define the problem and its solution/s. To the degree that such collaborations actually occur, there should be some evidence that all stakeholders entering the relationship are, in fact, endowed with some perceptible attributes applicable to the problem at hand. This chapter goes beyond the activities of the case studies themselves to identify those attributes that allowed the individual NGOs to be considered by their GO counterparts as potentially worthwhile partners, and compares the mechanisms by which the collaborative alliances were established. As the narratives in the previous chapter indicate, each of the NGOs studied was successful, to varying degrees, in creating or maintaining constructive relationships with government agencies although, as will be discussed, these may not all be accurately described as relations of critical collaboration. If an examination of these several cases can give an indication that a variety of organizational factors do lead to collaborative relationships, the case can be made that critical collaborations can occur in a variety of settings, between organizations with diverse characteristics.

Having said this, what can be learned about critical collaboration from these five case studies? To what extent is each of the cases an example of this form of association? It is premature to make a final decision in the sense that the projects under study continue to evolve, but the activities of both NGOs and GOs to date may give an indication of the degree to which each case has evidenced collaborative activity between the NGO and GO, and the expected direction that these relationships may take in the future. Whether each of the parties involved in the ensuing relationships can be seen as being involved in a "critical" collaboration (as defined in previous chapters) will also be explored in this chapter.

Overview

To what extent can the cases be viewed as examples of critical collaboration? It is difficult to identify some threshold in an NGO-GO relationship, beyond which it can be recognized as such. First, it is important to understand that critical collaboration is a process, rather than a goal or a state of organization. Next, each collaboration in the study is distinct in the way it was established and the degree to which the relationship has progressed. For example, PRADAN—a large, well-known NGO with a proven record of collaboration with GOs throughout India—made the request to participate in the design and implementation of the tank rehabilitation project. Auroville's watershed project, in contrast, grew out of what was perceived as a growing crisis by GOs, international donors and AWS itself. PHCC initiated its relationship through sharp critique of government actions and inactions, while SPEECH's relationship with a state GO grew from an outside donor's requirement that its funding of the program would be conditional upon the participation of the rural beneficiaries in program/project design and implementation. In addition, the experience of SHIELD indicates a feature of the shared jurisdiction over rural development that is indicative of Indian federalism, in that the NGO was able to find financial support for a small-scale afforestation program from an agency of the Central government after being denied support from the state's forest department.

One measure of the existence of a critically collaborative relationship, though, might be inferred from a formal declaration of the parties to work as partners. This reflects an important feature of the concept: that the decision to work together as *critical* partners, as distinct from other, less interdependent relationships, is a deliberate decision on the part of all stakeholders. In the cases of PRADAN and PHCC, each NGO signed a Memorandum of Understanding (MoU) with its GO counterpart as a means of formalizing the collaborative roles. SHIELD's relationship with the National Afforestation and Eco-Development Board (NAEB) was established through the approval of a grant application, and confirmed through an evaluation completed for NAEB by an outside consulting firm. SPEECH entered into contracts with individual GOs for PRA training. At the time of the fieldwork, AWS had not signed any formal agreements with its associated GOs, yet this can be anticipated, as the MoU is standard procedure for NGO-GO interactions.

The extent to which the relationship between NGO and GO in each case can actually be considered collaborative varies (see Table 6.1). In all the cases there is a common factor, in that each has moved beyond the original point of contact at which the two organization initially agreed to work together, and all have continued for at least several years. This is not to suggest, however, that the future existence of the relationship is guaranteed. PHCC's ongoing relationships with various local GOs indicate that the collaborative process is a dynamic one, and that the degree to which the partners choose to continue those relationships varies in each instance.

Table 6.1
Perceived Degree of Critical Collaboration among the 5 Case Studies*

Linkage	Cooperation	Operational Collaboration	Critical Collaboration
		PHCC ←→	
			PRADAN →
		AWS →	
		SHIELD →	SPEECH

* Arrows indicate expected future level of collaboration

Both SHIELD and SPEECH are involved in programs that are far less complex in design than those of either PRADAN or AWS, and required little negotiation regarding the particular policy aspects involved. SPEECH's experience indicates that while the NGO-GO relationship has not become more complex in terms of a greater interdependence between the NGO and any specific GO (in terms of 'scaling up' the NGO's influence), the NGO has "scaled out" its operations through the creation of a new organization to teach PRA techniques to an increasing number of GOs in the state. The study of AWS suggests the existence of an incipient, yet growing, case of collaboration. And, in the case of SHIELD working with NAEB (and, to a lesser degree the experience of SPEECH), the question arises as to the extent to which the relationships are really examples of *critical* collaboration, or if their relationships fall short of the definition.

The factors investigated to test the validity of the central question—the attributes of the NGOs and the GOs involved in the cases studied, as well as the contexts that shape their working relationships—were outlined in Chapter 1. These also appear in Table 6.2, which highlights the variations found in each of the factors identified as important to the creation of critical collaborations. During the interviews, some of the indicators were seen as central to the relationships, while others were perceived as less important than originally expected. Several indicators also appear to have been of greater importance to particular cases, while relatively unimportant to others. Moreover, as is the case when studying any complex system, it is difficult to disaggregate the effects of a single factor from the entire set of circumstances, of which each is an interconnected part.

Table 6.2
Factors Leading to Critical Collaboration

NGO	Size	Approach to Government	Initiator of Contact	Source of Capacity	Source of Legitimacy
PRADAN	L	Mainstream	GO request	T, Ex	History
AWS	L	Mainstream	OutsideDonor	T, Ex	History
PHCC	M	Social Action	Advocacy	T, Ad	History
SPEECH	S	Mainstream	OutsideDonor	SK	SK
SHIELD	S	Mainstream	NGO request	Association	Association

Legend: NGO Size: S = small   M = medium   L = large
Source of Capacity/Legitimacy: T = technical   Ex = experience
Ad = advocacy
SK = specialized knowledge   Association = NGO-NGO association

For instance, the local policy culture (LPC), as one of the indicators of the context in which collaboration takes place, was not seen by the NGO or GO officials directly as a determining factor in the collaborations *per se*. This is not to suggest, however, that the concept of the local policy culture is not a valid one, or that local factors did not influence any of the projects. Several cases indicate that individuals at the local level were influential, if not instrumental, in the creation of the NGO-GO relationship.

The issue of the NGOs' ideologies—their ideas of the roles they play in conducting rural development—seems to be consistent among the NGOs studied. They view their primary activities as: 1) empowering their rural constituents to, eventually, have the means to confront state and local GOs on their own; and 2) acting as intermediates between GOs and the poor. There was no recognition on the part of the NGO personnel interviewed that the two cannot be conducted simultaneously. The only possible exception to this occurred in the case of AWS, which appears not to have had any clear idea of their role regarding the surrounding villages at the initial stages of the watershed program, although the realization of this shortcoming has led to a conscious effort on the part of the NGO to learn what they can do as an organization to include the local populace in the design and implementation of the program.

How those ideologies are translated into action is another issue. The *modi operandi*, the manner in which the GOs and NGOs worked together, appear to be functions of the individual NGO's ability to constructively engage the government. In two of the cases—PRADAN and AWS—the NGOs had relatively little difficulty in directly engaging the corresponding GOs; in both cases, their participation was easily initiated. At first view, this might appear to be related to NGO size; both PRADAN and AWS are highly visible and influential organizations, and were sought by GOs for their participation.

There were no instances of an NGO having strong objections to working with GOs (although there were several other NGO directors interviewed who steadfastly refused to have any contact with GOs or bureaucrats). PRADAN, though, is the only NGO of those studied that has a well-established history of actively seeking partnerships with GOs, as its original mission statement indicates (although, as with its decision to participate in the state's irrigation tank program, PRADAN carefully chooses the programs in which it will participate). SPEECH, on the other hand, was initially wary about working with GOs, yet not unwilling to serve as a trainer. AWS sees partnerships with GOs as necessary to the accomplishment of its objectives, yet does not appear to be overly sanguine about the process. PHCC, despite its aggressive approach to GOs in the defense of the local natural environment, favors working cooperatively and collaboratively with GOs, although it sees no contradiction in taking organizations and individuals, even those with which they are collaborating, to task for issues that it sees as important. Even after almost a decade of challenging GOs in the Kodaikanal area, PHCC continues to see one of its most important roles as a critic and watchdog of government activities, while at the same time it works closely with GOs to solve problems. SHIELD, the smallest of the NGOs studied, has had little actual contact with the Central government other than providing documentation of its work (which is channeled through an intermediary organization).

The question of technical expertise, one of the proposed indicators of NGOs, is one aspect of a larger issue, that of capacity. The issue of organizational capacity will be discussed below, along with the associated factor of legitimacy.

The issue of NGO-NGO relationships was a major factor in two of the cases. Officials of both AWS and SHIELD indicated that it was necessary for the NGOs to look outside their own organizations for the resources to give them the requisite capabilities to work with government. In the case of AWS, the NGO's own self-perception as a legitimate NGO—a component of their strategic competence—was lacking, and has become a part of a learning process included in its 'General Strategy' toward the planning and implementation of the watershed rehabilitation program. In the case of SHIELD and its relationship with the NAEB, the directors of both SHIELD and PRADAN indicated that the association between the two, and the support given by PRADAN to the smaller organization, were central to the NAEB's approval of the smaller NGO's afforestation project. While these two cases involved specific interactions with other NGOs, all of the cases indicate that relationships between the NGO and outside individuals and/or organizations have been important, if not vital, to their successful operations, in terms of either establishing or reinforcing their legitimacy and capacity vis-a-vis their GO counterparts. The incidence of NGO-NGO cooperation elsewhere in India is becoming an important feature of rural development (see Box 6.1).

Box 6.1
Federation of Voluntary Organisations for Rural Development—Karnataka

Since the early 1980s, a growing phenomenon has been occurring in the south Indian state of Karnataka: rural development NGOs of various sizes in that state have been joining together to form networks and coalitions, both at the district and state level. The largest and most influential of these, the Federation of Voluntary Organisations for Rural Development–Karnataka (FEVORD-K) is a good example of NGO-NGO cooperation.

FEVORD-K was created in January 1982, following a series of exploratory meetings held by interested NGOs. Since then, its membership has grown, from twenty-three (in 1983) to more than hundred and ninety today, and currently operates on a budget of approximately Rs 650,000 (Ananda 2001). It is an unusual organization. It is a loose connection of NGOs, with only four full-time staff members. The organization does not involve itself with the internal functions of its members, nor does it evaluate or fund them. It has no projects of its own. David Potter (1995: 3) writes that "the life of FEVORD-K is primarily in the countryside amongst the members. If the members are not singly or jointly active, FEVORD-K is not active."

The objectives of FEVORD-K are:

- to promote and develop areas of mutual help, cooperation, and understanding among member NGOs;
- to strengthen existing agencies and to foster new initiatives;
- to organize training programs appropriate to member organizations;
- to disseminate information; and
- to establish rapport with governments, donor agencies, and other sectors.

Like PHCC (see Chapter 5), FEVORD-K participates in both "outside" (confrontational) and "inside" (cooperative and collaborative) activities with various governments and agencies. Of the former, the most visible have been a campaign against the proposed GoI code of conduct of 1985 (Ramaswamy and Prasad 1990), and a protracted struggle (1984–1993) to overturn an agreement between the GoK and a private corporation to develop a pulpwood plantation on more than 30,000 ha of village forests and common lands (see Kanavalli 1993). At the same time, though, FEVORD-K has developed a number of venues to work cooperatively with the state government: the GoK has granted FEVORD-K representative status in its rural development committees, and a consultative group consisting of NGOs, the state's Development Commissioner, and heads of government departments has been functioning since 1984. The GoK has directed that all relevant government orders, circulars and policy papers be made available to FEVORD-K. Bhat and Satish (1993: 160) write that "the possibilities of influencing policy and decision-making have encouraged FEVORD-K to put substantial efforts into developing links with government, and the GoK response has been favourable."

Sources: Ramaswamy and Prasad 1990; Kanavalli 1993; Potter 1995; Rajasekhar and Nair 1995; Ananda 2001.

Other indicators were seen as important to the establishment of collaborations, both by the NGO and GO members involved, and in the course of the study of their collaborations. A review of these factors comprises the following discussion.

Conditions Necessary for Collaboration

Government Organizations and the Political Context

The political context within which governments and NGOs work is a major component of the milieu within which development interventions take place. The argument could be made that, to a significant degree, the ability of Indian NGOs to work together with government agencies in partnership derives from some distinctive features of the Indian political scene.

As mentioned in Chapter 2, India may well present a political atmosphere that is not readily apparent in other developing countries. Political dissent and protest are generally accepted tools for the rural poor, as well as other groups, classes, and castes to make their grievances known to the government. The system is relatively stable; there are many crosscutting social and cultural distinctions in Indian society which tend to compartmentalize and prevent widespread unrest. If, indeed, the Indian political system is endowed with the level of toleration and acceptance of public activism widely mentioned, one might expect that efforts on the part of NGOs to influence government policy toward rural development would have become a regular component of the political landscape, and that there may have emerged institutional arrangements that would allow, or even encourage, large numbers of Indian NGOs to participate in the rural development policy process. However, this has not occurred to any great extent.

Despite a sense of mistrust and jealousy on the part of many GOs toward other actors, the attitude of the GoI (as well as of some state and local GOs) has shifted toward a policy of fostering a greater degree of participation on the part of local governments, the population, and NGOs in the development process. This has led, at least at the level of national policy-makers, to a greater sense of cooperation than of hostility. The concepts of decentralization of decision-making and planning of rural development have been a recurring concern of the Central government,[1] and the GoI has adopted an official policy of encouraging NGOs to strengthen the ability of the rural poor to demand their legal rights from government. And, as has been mentioned, the creation of central ministries and agencies to work directly with NGOs at the grassroots level—either bypassing, complementing or working directly with state and local GOs in an effort to overcome the inertia that often brings local efforts to a standstill—has blurred constitutional distinctions between the roles of the Central government and the states in the policy process. The GoI has taken advantage of the federal system of government in assuming shared jurisdiction for some environmental fora that had previously been limited to the states by the Eleventh Schedule of the Constitution.

The level of active participation of the Central government in local NGO activities is evident in several case studies. In the case of PHCC, the National Wasteland Development Board (NWDB) worked with the District Collector to provide funding

[1] L.C. Jain cites numerous government efforts to decentralize development in India, beginning with the creation of a Royal Commission Upon Decentralization that operated from 1907–1909 (Jain 1985).

and municipally-owned land for PHCC to initiate their program. The National Afforestation and Eco-Development Board (NAEB) provided funding for SHIELD's afforestation efforts, which had been rejected by the state's Forest Department. It should also be noted that several district-level agencies (notably the office of the District Collector and the District Rural Development Agencies [DRDA]) are GoI bureaucracies, and staffed by IAS officials.

The lack of coordination between policy decisions made at the national level and the implementation of the same at the local level, however, remains a major obstacle. Despite the GoI's recurrent efforts to increase local participation in rural development planning, it has been largely unable to mandate cooperation by individual states, block- and district-level bureaucrats, or local *netas* (elites and power brokers). There is no certainty that changes in central policies will be transmitted successfully to the state and local levels, even for larger NGOs with the levels of funding and legitimacy enabling them to engage policy-makers at the national seats of government. A PRADAN document echoes this concerns:

> This is in fact, the crux of the problem; viz, how to establish relations with the lower orders of bureaucracy and the NGOs. There has never been a significant problems with the officials who are secretaries to the government and the NGOs. They are normally accessible to staff from NGOs and provide them the necessary support. However, they have never been able to transmit their enthusiasm for the NGOs lower down (PRADAN 1996c: 23; see also Jaix 1992).

At the local level, collaboration (or even cooperation) between NGOs and GOs remains highly dependent on the individual agency or decision-maker, despite national and state policies that encourage and mandate greater decentralization of, and local participation in, natural resource management. Lele (2000) points out that GOs are often willing to allow the participation of actors outside their own agencies only when the resource in question is in poor physical condition (joint forest management programs are generally limited to degraded forests or wastelands), or when the resource is socially or logistically difficult for the department itself to manage (such as small, "village" water tanks). This factor may account, to some degree, for the difficulties that PHCC and SHIELD encountered in working with the TNFD, as well as PRADAN's ability to redesign some of the government's policies toward tank management.

Government-NGO relations often depend upon the personal interests of the District Collector, who is a central actor for many rural development activities. The director of AWS pointed out that the organization (as well as Auroville as a whole) has been generally more willing to work with the government of Pondicherry than that of the state, for several reasons: in part due to the Territory's smaller size and the ease and frequency which the GO and NGO interact (due in part to the geographic proximity of the Pondicherry government offices), which facilitates personal interactions, and because the District Collector of Pondicherry (at the time) had an ongoing interest in the activities in the township and was disposed to assist their efforts (Giordano 1997). SPEECH's linkage with the state Agriculture Engineering Department in teaching PRA techniques was enhanced by the interest of an individual IAS officer detailed to work with DANIDA as a liaison to local GOs in Tamil Nadu (Venkatasamy 1997).

As mentioned in the previous chapter, SPEECH has been sought by a number of Tamil Nadu GOs for PRA training. The Revenue Department, seen by many as the most monopolistic, intransigent and powerful of the state agencies, displayed an uncharacteristic willingness to collaborate with a small NGO. According to an IAS official, the Director of the department wanted to have his staff trained in PRA in order to change their approach to rural clients, "Ordinarily, when people from his agency go to the village, they want the people to fear them; this is interpreted as a show of respect. The Revenue Department's director wanted to use PRA to foster the trust of the masses" (Venkatasamy 1997). Many GOs, however, continue to resist the participation of outside organizations and individuals in what they deem to be their jurisdiction.

Another important factor in the operation of all organizations is funding. Organizations such as PRADAN, with diverse sources of funds, have the ability to operate in ways that smaller NGOs cannot. For example, in addition to conducting their own experiments and trials without having to receive permission for its activities from GOs or outside funding agencies, PRADAN was able, in the case of the tank program, to continue to finance their operations temporarily while bargaining for modifications to the schedule of payments to workers. The ability to initiate and maintain operations without waiting for the appropriation of government funds (which may entail a lengthy bargaining process) may be crucial to operations, especially if a program is sensitive to seasonal climatic changes. NGOs that depend on governments for large portions of their operating budgets, however, are much more likely to be required to conduct their activities in accordance with government oversight and limitations.

Sources of program/project funding also affect the nature of the relationship between NGO and GO, as well as those between the two and outside donors. Foreign donor organizations frequently place conditions on the funding of a project, or make suggestions concerning the content or focus of the program, which may be the initial impetus behind a working relationship. In the case of SPEECH it was, in part, DANIDA's emphasis on the state Forest Department's use of PRA techniques in their social forestry program that made it possible for such a small, geographically isolated NGO to play a significant role in a GO-led program. Similarly, PHCC became involved with the state's Forest Department in the Kadavakurichi Hill program because of the involvement of the Swedish International Development Agency (SIDA) with the Forest Department and their influence in bringing the two organizations together. The initial withholding of funding for the watershed program of AWS was, in part, due to the Canadian government's concern that the original program plan placed too much emphasis on the engineering aspects of the work and not enough on the needs of the local villagers.

As noted, the associations of smaller NGOs with larger, well-established organizations—both indigenous NGOs and international organizations—appear to have been crucial to their ability to work with government agencies. Just as foreign donors can serve to enhance the positions of small, resource-poor NGOs, associations with other NGOs can play a significant role in enhancing a small NGO's ability to contribute to a collaborative relationship. At the same time, though, donors themselves often have a strong voice in shaping the directions that rural development interventions take, or place demands on both GOs and NGOs to perform in accordance with their requirements or face forfeiture of the their funding.

These factors all point to the idiosyncratic nature of NGO-GO relations in Tamil Nadu and the fluidity of the contexts within which collaborations occur. Both GOs and NGOs vary according to their individual characteristics, yet there appears to be a greater degree of variation among NGOs. The review of the cases now turns to an analysis of some of the attributes of the NGOs that made it possible for them to establish collaborative relationships with government.

Attributes of the NGOs Studied

From the evidence presented in the research, it appears that a number of substantial differences exist between the NGOs under study. Are those differences incommensurable, or is it rather the case that they instead engender varied, yet effective approaches toward similar outcomes with local and state governments—different paths to reach the same goal? As seen in the previous chapter, the NGOs studied have certain qualities that made them credible partners for collaborative relationships. As has been suggested, these qualities are reflected as power—"the power to define the problem and to propose a solution [which] is effectively shared among the decision makers" (Gray 1989: 119). What features do these NGOs possess that make them viable partners to government? Are there any apparent characteristics of the NGOs studied which might contribute to an explanation of their ability to create (or maintain) collaborative relationships with GOs in Tamil Nadu? Size is the most frequently invoked criterion for NGOs' ability to engage government, and it is to this indicator that the discussion first turns.

As mentioned in Chapter 3, the majority of indigenous NGOs in developing countries are small and resource-poor, and thus operate within a limited geographic area. Only a small proportion of a country's NGO sector is "visible" to their own Central governments as well as outsiders, and these few large NGOs tend to have a disproportionate impact on both domestic and international policy-makers. An NGO's size may affect its ability to perform certain roles. Larger organizations can be involved in several, sometimes unrelated, activities simultaneously, while smaller ones are more limited in their scope and variety of activities. Larger NGOs are considered to be able to reach policy- and decision-makers at the central level and affect central policies in ways that are not available to smaller groups (Bragg 1996). In addition, larger organizations are more likely than small, financially limited ones to spend their time and resources to reflect about who they are, what they are doing, and to plan for future operations and changes.

Size, along with the other attributes that accompany it, does make a large NGO more visible to governments and to funding sources, and may enhance the NGO's ability to make its voice heard at the centers of national policy-making to a much greater degree than the majority of smaller groups. However, this does not necessarily translate into improved relations with GOs at the local level, where policies are interpreted and implemented. Nor does it guarantee that an NGO will possess the capacity to engage in partnerships with other organizations (see Box 6.2).

Box 6.2
The NPIC in Gujarat

In December 1983 a newly created GoI agency, the Department of Non-Conventional Energy Sources (DNES), inaugurated a nationwide effort to distribute technologically improved *chulhas* (cookstoves) under the National Program for Improved Chulhas (NPIC).

Soon after the program began, officials of the new agency realized that they lacked the organizational and administrative resources necessary to implement a national effort, and turned to large NGOs for assistance in

(Box 6.2 contd)

(Box 6.2 contd)

the implementation of the program. The effort foundered, and DNES again restructured the program, to include the participation of state agencies in identifying and recruiting local NGOs for village-level training and for the construction of chulhas.

In the state of Gujarat, the Gujarat Energy Development Agency (GEDA) was chosen to administer the program. GEDA selected twenty-five local NGOs to implement the program. According to Maniates (1990: 108) GEDA chose NGOs on two criteria. First, it sought "trustworthy" NGOs (according to the author, trustworthiness "appears to be a function of the degree to which the NGO looks and behaves like the technocratic GEDA"). The second criterion was the NGOs' ability to produce timely documentation of their activities (this reinforces Khator's discussion of the "pretentious policy approach"—see Chapter 2).

Several of the NGOs initially chosen to participate in the NPIC were relatively large organizations with well-established technical capacities, currently conducting multiple programs in several communities. It was expected that these NGOs would be ideal collaborative partners. As it turned out, though, a number of the organizations

> found that the costs of collaboration to be too high. Collaboration with the government required a commitment of additional resources by the NGO....Frequently, the organizational cost of meeting these demands was perceived to be too high and, thus, NGOs with active programmes in other areas often declined to participate in the cookstove programme.

Ultimately, smaller, less experienced NGOs (with limited technical expertise and few ongoing programs) were found to be more appropriate partners for participation in the NPIC, primarily because they had the time and interest in creating and maintaining a collaboration with GEDA.
Source: Maniates 1990, 1992.

For the two largest NGOs studied here, size alone has not been an automatic determinant of the successful achievement of their goals. In the case of PRADAN, its size has given the members access to directly appeal to central and state government agencies for authorization to change rules and regulations. However, as one observer notes, this clout has meant that PRADAN, as a result of bypassing local officials and GOs in favor of working with state or central-level governments, has raised the ire of a number of local bureaucrats who resent the imposition of higher levels of government into their areas of jurisdiction. A former District Collector for the Madurai District noted that some local-level officials dislike working with PRADAN because of their perception that the NGO 'bypasses proper channels' and, instead, appeals directly to state and central government agencies. Because PRADAN has such power to contact and influence higher-level policy makers, it tends to overlook the input of local officials in the creation and implementation of projects and instead impose its own decisions and procedures upon local agencies (Rajivan 1997).

The Auroville Water Service faces a different problem. It is not so much that their size enhances or limits their effectiveness, but that they have not, in the past, had a real sense of mission regarding the villages surrounding them. In terms of access to funding, technical expertise, and other resources, the functional units of Auroville have the means to conduct large-scale development activities. However, AWS lacked of a basic vision of their role as an agent of development outside their own borders, as well as actual experience in working *with* the local villages (as opposed to hiring villagers to work *for* them in their own activities); the director realized that the scope of the problem would require that AWS reorient itself and its mission to include the participation of local population in the watershed program. Despite Auroville's size, resources, endowments, international connections and wealth of technical expertise, its constituent working units have had little experience of working with groups outside their own borders, and no formal mechanisms exist by which such experiences can be shared internally. Auroville was successful in creating its own, somewhat insular, social environment, which has been seen by many to be insufficient for their existence in a larger context. AWS is working to overcome this perceived shortcoming.

Coordination with multiple agencies and levels of government is what the AWS sees as potentially the program's greatest obstacle. Since they are designing a multidimensional program, the components of which address varied aspects of the natural environment (including flora and fauna restoration, afforestation, surface- and groundwater management, and agriculture) the AWS will, by necessity, be involved with a number of district, state and central bureaus, agencies, and ministries under whose jurisdiction these components fall. Moreover, many of these issues are within the jurisdiction of more than one agency. The process of successfully coordinating with not one, but a number of GOs (at local, state, and central levels), whose interrelations with each other (as well as with NGOs) are often less than harmonious, makes implementation considerably more difficult (Giordano 1997).

Auroville's relationship with government parallels those of other large NGOs in regard to its access to resources, especially foreign funding. Its large size and proven history of accomplishments makes it attractive to GOs looking for partners. Auroville has an additional advantage when dealing with government officials: because of the number of expatriate experts who work there, and their connections with foreign governments, NGOs and funding agencies, Auroville members enjoy a cachet denied most Indian NGO officials. According to Thomas Gablier (1997), a founder and the current director of AWS,

> When a government official finds out that it is someone from Auroville, the doors open for us. We are much more likely to be allowed to have our case heard, and can generally bypass the usual lower-level red tape—although this does not automatically mean that we are more likely to get a favorable answer.

As the government came to see that the AWS was not out to steal their authority, and they "learned that they had nothing to lose," GOs became more willing to entertain the possibility of policy change (Gablier 1997).

For the two largest NGOs, then, the factor of size—in and of itself—is neither a constraint nor an advantage. Both have substantial and multiple sources of funding, a diversity of activities, the available assistance and support of many outside organizations, and a broad network of both indigenous and international supporters. For both, access to government officials and policy-makers at multiple levels is not seen as a constraint. Nevertheless, the consequences of the two largest NGOs' size may actually work against them. PRADAN's access to the centers of policy-making—at the expense of working more closely at the local level to design rural development interventions—has resulted in a level of mistrust of local bureaucrats which, in turn, may compromise better relations at that level. Similarly, Auroville's clout—including the international attention (and funding) it has received since its creation in the late 1960s, as well as its concentration on its own growth and well-being during that time—led to a sense of mistrust among both GOs and villagers, who see Auroville as "pushy and arrogant" despite the widespread view among the local population of Auroville as a "transparent" and reliable entity (Giordano 1997).

Closer to the other end of the size continuum, PHCC, SHIELD, and SPEECH have indicated through their experience that small organizations can, under propitious circumstances, interact positively with government agencies. However, the ways that such interactions are initiated differ distinctly from the manner of both PRADAN and AWS, and from each other as well. The cases nevertheless indicate that they have been able to work with government agencies in ways that are proportionally greater than considerations of their size alone might be expected to allow.

SHIELD was unsuccessful in working with state-level bureaucrats but was able, with the backing of a larger, well-established NGO (PRADAN), to work successfully with an agency of the Central government. SPEECH, because of their possession of a set of skills needed by government agencies— the ability to conduct and teach Participatory Rural Appraisal (PRA) techniques—was sought by a number of GOs for training purposes (although the interest on the part of many of the GOs was the result of donor organizations' imperatives). In addition, it appears that PHCC, through its campaign of public advocacy, actually created changes in the political environment which, in turn, enabled the NGO to work with government: they mobilized the support of local populations to create highly visible appeals to established environmental and land-use laws, along with a body of technical data and environmental experts, to back their assertions and advocacy.

If size were the sole determining factor in choosing an NGO for inclusion in a collaborative partnership, it might be expected that the smaller NGOs would have been far less successful than the larger in establishing such relationships. Yet all have been able to enter into some form of cooperative relationship, several of which may qualify as critical collaborations. What other factors are important to the creation of collaborative relationships?

One means of judging the suitability of an organization to work with is to review its past record of activities and accomplishments. GOs are more likely to choose to work with an NGO whose "track record"—its history of attempts and accomplishments, its approach to development, and past working relations with government—indicates the possibility of a successful partnership. Less experienced and, in many cases, smaller NGOs have a more limited history of activities and accomplishments, and may have no experience working with outside organizations. It is less likely that GOs will choose to work with NGOs about which they have little or no knowledge.

This feature—the organization's "track record"—was seen by both PRADAN and AWS as an important attribute in the establishment of their relationships with government.[2] Both had built their reputations in a different manner. PRADAN, through continued, deliberate efforts to work in cooperation government agencies at several levels, had built a record of solid accomplishments in the field of rural development. Auroville (of which AWS is a constituent organization) has a history of more than 30 years of accomplishments upon which it could stand: a long-term presence in the area, proven achievements, and widespread international connections (although these were not accomplished in the context of rural development *per se*). Both NGOs were "known quantities" in regard to their accomplishments.

---

[2] The concept of an NGO's "track record" is also viewed by government as an important factor. Potter (1995: 25–27) writes that both a 1990 GoI Government Order and a GoK policy (released in 1993) explicitly directed state Forests Secretaries to give NGOs "with proven track records" positions in their Joint Forestry Planning and Management (JFPM) programs—although, according to Potter, "not a very central one."

In comparison to the larger, more established groups, the smaller NGOs lacked extensive records of concrete accomplishments; this makes GOs' determinations of their suitability difficult. In the absence of such a record, these NGOs were still able to establish relationships with GOs through other means.

The director of SHIELD attributed that organization's ability to receive funding from the NAEB to their association with, and support of, a large, well-established NGO in the area—PRADAN. In this case, the support of a larger, well-established NGO served as a guarantor for the smaller group's performance, especially in light of SHIELD's lack of past accomplishments (this factor was stressed by the directors of both SHIELD and PRADAN; both emphasized that without the association, SHIELD would have had little likelihood of obtaining a grant from NAEB).

In the case of SPEECH, the relationship between NGO and GOs resulted from the actions of an IAS official who was interested in learning how PRA—which the NGO had been trained to use—was actually employed by a "practically-oriented" NGO (Venkatasamy 1997). Subsequently, the official, who was acting as a liaison between a foreign donor and the state Agricultural Ministry, initiated the NGO's work with state GOs when he requested the services of SPEECH in training the GO in the use of PRA. Neither age nor history of the NGO were mentioned as relevant considerations.

PHCC was able to overcome the handicap of being a less well-established NGO through its vociferous public advocacy campaign. A combined effort of ongoing activities, according to Jeyakaran (1996, 1997), gradually convinced GOs in the Kodaikanal area to accept the role of PHCC as a working partner. One major factor in the emergence of an increasingly amicable working relationship, according to one of the founders of the organization, was that the period of several years of vociferous public confrontation was accompanied by PHCC's constructive environmental efforts (including a number of detailed natural resource studies, technical assessments and parallel afforestation projects being undertaken simultaneously). Over time, PHCC's actions regarding the protection of the area's natural environment were perceived by a number of GO officials as sincere and politically disinterested, and the NGO was eventually successful in creating a sense of legitimacy for itself as an actor in local environmental issues. This, in part, led to the formation of cooperative and collaborative relationships with a number of local GOs. PHCC was seen (by GO officials) not as a "flash in the pan", but as a genuinely concerned group of citizens whose advocacy was supported by concrete action. Knowledge of the NGO's past activities assuaged many officials' fears that PHCC could be a possible threat to local politicians (Jeyakaran 1996).

An organization's track record (or lack thereof) was also seen as important in the case of the AWS. Despite Auroville's history of restoring the health of the area's natural environment, the Water Service's lack of orientation toward the human dimensions of rural development was clearly a limiting factor in their ability to receive funding. Also, the failure on the part of AWS to encourage local villagers' participation from the outset of the program's planning stages (although enthusiasm for the program on the part of some villages and villagers existed) was seen by the director and others as a threat to the program's success.

Findings: Capacity and Legitimacy

The results of the case studies indicate that, underlying the various characteristics of the organizations and contexts within which the relationships take place, there are two basic issues related to the success of creating a collaborative relationship, both of which are perceptions (held by other stakeholders) of a potential partner's suitability as a working partner. The first is *legitimacy*, the degree to which the organization is seen as having a credible interest, or "stake" in the situation at hand; the second, *capacity*, is the perception that the organization can participate as expected. Both attributes are important to the establishment of critical collaborations.

The aspect of actors' legitimacy as stakeholders is a necessary, although not sufficient, condition on which to base a collaborative relationship. As mentioned in Chapter 4, the burgeoning number of Indian NGOs created in the past two decades has led to an ongoing controversy concerning their legitimacy as a sector, including the debate over the adoption of an NGO code of conduct. On the other hand, governments in India are widely viewed in general as legitimate actors in the rural development process.[3] This does not mean, however, that GOs automatically possess the capacities required for participation in collaborative relationships (see Box 6.3).

Box 6.3
MYRADA and the PIDOW Project in the state of Karnataka

MYRADA is a large NGO headquartered in Bangalore, with projects located in the south Indian states of Karnataka, Tamil Nadu, and Andhra Pradesh, and supports projects in several other states, and in Myanmar. Like PRADAN (see Chapter 5), MYRADA is involved in a wide range of rural development activities, and is funded by a number of both indigenous government and private sources and international donors. With a staff of four hundred and sixty and more than fifteen hundred volunteers MYRADA, created in 1968, has made a conscious decision to seek collaborative working relationships with government agencies. However, as its director writes,

[w]hile MYRADA does collaborate with government in implementing ...programmes it has been careful to avoid the image of being a "turnkey" operator or a contractor. If this image of a contractor is allowed to grow, MYRADA would lose its flexibility and the ability to press for change where it is necessary (Fernandez and Mascarenhas 1993: 154).

In 1984, the Government of Karnataka (GoK) signed an agreement with the Swiss Development Co-operation (SDC) for an integrated watershed development program, the Participative Integrated Development of Watersheds (PIDOW). The program was designed to incorporate the physical

(Box 6.3 contd)

---

[3] Ranjit Gupta (1986: 329), in an article comparing the performance of two dissimilar NGOs in south India, notes that "rural development across the country on a scale commensurate with the size of the problem, cannot be undertaken by any agency *except* the government." In particular reference to India, John P. Lewis (1962: 26) wrote that "the proposition that it is the business of government to be the principal planner, energizer, promoter, and director of the accelerated development effort...is so fundamental and so little disputed in India that one would probably not bother even to mention it to an Indian audience."

(Box 6.3 contd)

> inputs and technical advice of a number of GOs. At the same time, SDC signed an agreement with MYRADA for the latter to assess the planned interventions, to facilitate popular participation in the program, and to test and develop new and alternative approaches and technologies for sustained participation and watershed management. According to Bhat and Satish (1993: 163), the state "felt that peoples' views on, and participation in, watershed development could be more effective if they were organized by NGOs."
>
> It was soon discovered that many of the line departments, due to their individual schedules and program constraints, were unable to participate in PIDOW. Subsequently, the program was placed under the aegis of an agency with previous experience in integrated programs, the district's Dryland Development Board (DLDB) [Oct. 1990].
>
> The case indicates that the GoK's approach to collaboration in the PIDOW case was innovative, in that it brought together a multidisciplinary team, composed of state forestry, agricultural, and horticultural departments and an NGO to create agreed-upon integrated plans for each micro-watershed involved. It also points out that GOs, as well as NGOs, do not necessarily possess the capacity to participate in projects and programs.
>
> Sources: Fernandez and Mascarenhas 1993; Bhat and Satish 1993.

The second factor, which involves issues of an organization's credibility (and is integral to perceptions of its legitimacy), is that of the stakeholder's *capacity* to perform: that in order to be seen as a credible participant in a collaborative relationship, a group or organization must "possess resources and skills sufficient to justify their involvement in collaborative efforts" (Gray 1985: 922). Both factors must be present. An organization seen to have a legitimate interest in a problem domain, but lacking the capacity to support its concerns with practical action, is unlikely to be successful as a partner. On the other hand, the organization that has capacity without legitimacy may be considered ineffectual or even harmful. Both factors were found to be central to the establishment of the relationships between the NGOs and GOs studied. In each of the situations discussed in the previous chapter, it was found that each NGO under study evinced a measure of both, although each did so in different ways.

Smaller NGOs (just as the rural poor for which they work) frequently are perceived to lack the capacity to be accepted as partners in government programs, even though they may be seen as legitimate representatives of the rural poor with whom they work. With limited tangible resources, and little visible practical experience in development implementation, their ability to engage government agencies in collaboration is constrained. How do such NGOs overcome what appear to be such constraints?

SPEECH, one of the smallest NGOs, was successful in working with local GOs primarily due to their possession of a specialized skill that was sought by a number of GOs. Despite other limited resources, the NGO had been trained in the use of PRA, which (as mentioned above) is a set of techniques increasingly required by donor agencies in their programs. This necessity, along with the intervention of an IAS officer, has led to the establishment of relationships between SPEECH and local GOs.

SHIELD, despite its inexperience and small size, was able to establish a cooperative, if not collaborative, relationship with a GoI agency largely due to two factors: a Central government agency that was willing to allow a degree of discretion to the local NGOs who received their grants, and the support of PRADAN for SHIELD as a capable actor.

It was necessary for PHCC to establish itself as a legitimate actor in the environmental concerns of the Kodaikanal area while, at the same time, demonstrating its technical capacity in order to make itself a more likely partner for GOs. The NGO used the tactic of public confrontation of local and state government as a means of bringing its concerns to the public agenda. Appeals to and demands on local and state governments to enforce existing legislation were used in the defense of local environmental conservation, as well as in the interests of the area's rural populations in the sustainable use of common property resources. Through its initially antagonistic stance toward the state government's tacit approval of illegal activities on the part of local builders and developers, and Kodaikanal's failure to enforce their own environmental laws, PHCC elevated the issues of local environmental protection and conservation of natural resources from non-entities to the level of widespread public debate; even the Chief Minister of the state was brought before the court for her involvement in the illegal hotel construction. The situation, in which the NGO in effect creates a forum to address environmental concerns where none had previously existed, is what Gray (1989: 123) refers to as "the power to mobilize" a problem domain, in which different groups having stakes in the economic development (and the repercussions) in the Kodaikanal area, especially those individuals and groups with no other effective voice, had the opportunity to air their concerns. As a result of its continued efforts, PHCC was eventually accepted by a number of GOs as a legitimate and concerned actor in the local environmental debate. Its capacity to participate with the Forest Department in the Kadavakurichi program was established through a history of achievements, its use of technical experts, and the presentation of a coherent plan bolstered by detailed data regarding the area and the problem.

PHCC's experience affirms an important point. The ability of an NGO to confront government action or inaction, or to engender popular protest on behalf of a particular cause, only results in attracting the momentary attention of the particular GO. Even if the NGO is seen as a legitimate representative of the poor, if it lacks something to back up its insistence with concrete alternatives, governments will do little more than to mollify those who challenge their authority. In order for the confrontational NGO—though it may be seen as a legitimate actor—to be perceived as more than a temporary nuisance, it must be ready to suggest what changes are necessary, and how such change can be realistically effected (Rajivan 1997).

In sum, the case studies indicate that there is no single indicator of the NGOs that can be identified as the ultimate causal factor in the creation of critical collaborations. Rather, each of the indicators recognized as being important to the establishment of the NGO-GO relationships was seen to contribute to perceptions on the part of the government agencies involved: that the NGOs possessed both the capacity and legitimacy necessary for the organizations to work together.

Conclusion

Having reviewed the individual cases for what are seen as important factors in their relationships with GOs in Tamil Nadu, the question should be posed: to what extent can these relationships be considered examples of critical collaboration? As defined in the beginning of the book, a critical collaboration is,

> one that allows both government and NGOs to work together as partners, or "stakeholders"—those parties having an interest in the outcome of a particular problem, and who are directly influenced by the actions taken by others to solve the problem—in both making and implementing efforts at rural development. The nature of this form of relationship is one in which an NGO has the capacity to participate in the decisions that shape the way that development interventions are conceived and planned, as well as the way they are conducted in individual circumstances.

It is evident that PRADAN possessed both the legitimacy and capacity necessary to be chosen as a collaborative partner in a large-scale government program. PRADAN, which has ample experience in working in a collaborative, as well as critical, manner with GOs, was sought by the state government as a partner in their tank irrigation program. As important as having both the legitimacy and the capacities necessary to perform with the relevant GOs is the fact that PRADAN has, as a part of their mission as an NGO, consciously and continuously worked to develop the capabilities to work with GOs in collaborative relationships.

The case of the AWS provides an insight into a program with perhaps the "youngest" NGO of all the cases. Despite its existence since the mid-1980s, AWS certainly has less experience—and, as a result, less legitimacy—as a working NGO (as they are defined in Chapter 2) than any of the other organizations, yet is a part of, by far, the largest, oldest and perhaps most influential of all the organizations studied here. Its ability to approach the creation a critically collaborative relationship with local governments appears to have been successful to date, yet its ability to accomplish its aims with local populations is still in question.

SHIELD has had little direct communication with the GoI, despite the fact that the two have entered into a relationship that allows the NGO a certain amount of discretion to shape the Central government's program to fit the NGO's (and the local populations') needs. While the NAEB's seeming laxity allows the NGO to tailor its program, the question arises: does the NGO have the ability to negotiate existing rules and regulations with its GoI counterpart? This cannot be answered at present. Currently, the relationship seems to fulfill the qualifications of a collaboration, but the extent to which the "critical" component of the relationship extends into the future may hinge on the GO's willingness to receive and respond to outside input, especially in what is seen as highly fluid political and organizational atmospheres.

SPEECH provides training for GOs in southern Tamil Nadu, yet participation on the part of some GOs has been less the result of an interest in collaborating than the requirement of outside funding agencies' exigencies: donors often make their funding conditional on the use of such methods. To the extent to which the GOs made the decision to improve relations with their rural constituents and subsequently agreed to accept training taught by the NGO, their relationships with SPEECH can perhaps be viewed to some degree, as critical collaborations.

In the situation of PHCC's collaboration with the Forest Department in the Kadavakurichi Hills program, it is evident that the NGO has, to date, been successful in engaging the FD critically. As such, the experience satisfies the criteria outlined for the identification of a relationship of critical collaboration. Whether this collaborative effort continues is not at all certain, given the GO's inflexibility toward the NGO's concerns. Without the conditionality placed on the state's Forest Department to work with PHCC by the program's donor, there is a strong possibility that the collaboration would not have occurred in the first place, and that the continuation of the relationship may hinge on this requirement.

A review of the cases indicates that the creation of collaborative relationships between NGOs and state and/or local government agencies was neither an isolated happenstance nor simply the extension of relatively beneficent feelings between a government agent or agency and a compliant NGO. Rather, the relationships that emerged reflect distinguishable and distinctive features of the several NGOs and the events leading to the establishment of their working relationships that occurred, indicating that similar outcomes may result from a variety of pre-existing factors. If this is indeed true, one must consider whether collaborations occur from some sort of convergence of favorable circumstances or, rather, whether there is an underlying set of preconditions common to collaborative relationships. Is it possible that, despite the seeming differences in the NGOs and the way that their relationships with GOs occurred, there is a common set of expectations that must be satisfied by the organizations involved in order for such collaborations to occur? Moreover, if such causal factors are common to a range of social and political contexts found within India, the question that arises is whether or not these factors are basic to the creation of collaborations in other political, geographic, cultural or social settings.

All the cases indicate the importance of NGOs possessing the qualities of both legitimacy and capacity. These are viewed as necessary, although perhaps not sufficient, conditions upon which the establishment of collaborations ultimately depends. If this is true, perhaps the two traits may be important indicators for the creation of collaborations in other countries. How GOs and NGOs in other countries, with differing political and social contexts, might be able to establish similar relationships will be discussed in the Afterword.

Chapter 7

Afterword: The Comparative Context of Collaboration

> The real question about collaboration is: do we have a choice? Given the long list of multiparty problems awaiting resolution, the record of success to date, the potential available, and the dire prospects of irreparable harm to our planet, can we as a global society afford not to search for common ground through collaboration?
>
> Barbara Gray (1989)

As shown in the Chapter 6, NGO-GO collaborations in Tamil Nadu are initiated under a variety of circumstances; the creation of such relationships is not determined by a precise set of preconditions. Collaboration in that state is, as elsewhere, largely a hit-and-miss proposition fraught with mistrust, individual caprice, competition rather than cooperation and is, too often, subject to the whims of those who hold and wield political and economic power over the lives and livelihoods of the rural poor. The purpose of this summary is to build on the findings in this study in order to identify some basic and replicable factors of NGOs and governments that underpin the creation of the critical collaborations. The factors are not necessarily applicable or feasible in every political or social context, but can be found outside Tamil Nadu and India, and should be replicable in a number of other areas.

Overcoming the Ideological Divide

One of the basic changes that must occur for GOs and NGOs to work together as partners is an ideological one. The rigid, artificial distinctions between the sectors—and the feeling that one sector of society is the best or the only legitimate actor in rural development—must, in some manner, be overcome.

Many scholars and practitioners assert that the artificial division of state, market, and civil sectors of society into discrete, compartmentalized entities creates an idealized vision of their roles in society. Realistically, the sectors are in truth more interdependent and interactive in the area of development than is generally acknowledged, and that these existing linkages should be recognized. The problems of development are far too complex to be amenable to one sector of society alone; there are caring and dedicated people in every portion of any society, and through their working together, novel and practical solutions to many problems can be found.

Similarly, the distinctions drawn between "top-down" and "bottom-up" approaches to development—the extreme positions that the knowledge and wisdom necessary to bring about development is the sole possession of either planners, technicians and university-trained experts (according to the top-down adherents) or that of the poor themselves (in the eyes of populist bottom-up advocates)—is overdrawn. Indeed, the rigid adherence to simplistic ideologies prevents those involved in development from realizing the possibilities of new and innovative approaches and policies.[1] Furthermore, as has

[1] Korten and Quizon (1995: 158) write that "The search for just, sustainable and inclusive approaches to development has suffered greatly from the dominance of ... opposing ideologies. If divorced from the reality of the prevailing human predicament, ideologies quickly move from useful action guides to becoming straightjackets that keep us from seeing important aspects of our reality that imperil our very survival."

been seen in this study (most notably in the case of the Tamil Nadu Forest Department), many GOs in India are loath to include outsiders from participating in the decision-making processes and management under their jurisdiction. How can those involved in rural development be assured that joining with other individuals and groups will be beneficial, and not harmful?

If both GOs and NGOs are to overcome their hesitations about working with each other in a mutually dependent manner they will have to have a more complete and realistic view of what they can expect of themselves and each other. This involves a level of reciprocal accountability and openness about mutual operations and expectations.

Creating a Level of Mutual Accountability

In Chapter 6, the concepts of legitimacy and capacity were identified as crucial to the establishment of collaborative relationships. The idea of inter-organizational collaboration is based on expectations of interdependent decision-making and performance on the part of all actors involved. The idea that a stakeholder can, has the ability to, and will successfully act as expected, is a basic requirement of all cooperative inter-organizational efforts. In order for both government agencies and NGOs to feel comfortable in offering and accepting collaborative roles, there must be a way that stakeholders can determine the characteristics, activities, and capabilities of those with whom they may wish to work. Especially if, as has been mentioned, the desired end of both GOs and NGOs is to scale out successful activities, each must be able to identify those organizations suitable for working in tandem, and be able to depend upon each others' fulfillment of the promises and expectations upon which the interrelationship is promulgated. At the same time, both GO and NGO leaders must be assured that the ground rules by which they will interact with government officials will be clearly elucidated and evenly applied. A clearly demarcated and level playing field—a stable, predictable, and transparent working environment—is a necessity, not only on the part of governments and NGOs, but donors, recipients, and all others affected by a given intervention. At present there is not a great deal of evidence that such an environment exists in India to any great extent. If collaborative relationships are to become more than occasional, haphazard occurrences, there must be some means by which the parties can reliably gauge each others' credibility and abilities. This requires, on the part of both GOs and NGOs, accountability to each other as well as to other stakeholders.

On their part, governments are generally characterized as having weak levels of accountability to other, outside entities for a number of reasons. In this regard, NGOs are viewed as appropriate representatives of the rural poor for the purpose of strengthening the society's ability to make the governments perform in accordance with their responsibilities (Farrington and Bebbington 1993).[2] The experiences of PHCC point out the role that NGOs can play in keeping GOs accountable to the public.

NGOs, lacking the legal and coercive power to maintain their presence and activities, face greater difficulties than do governments regarding questions of accountability. Since accountability and autonomy are basically antithetical, NGOs often face the possibility of losing a large measure of their independence when they are subject to outside scrutiny. As important as a degree of accountability may be to collaboration with government, many NGO officials feel that the maintenance of autonomy is essential to their operations, and that some aspects of autonomy must be carefully protected even as they move toward NGO-GO collaboration. Several authors suggest that NGOs may actually be able to maintain a degree

---

[2] The role of NGOs in acting to strengthen the poor through "consciousness-raising" is criticized by Hashemi (1996: 128), who asserts that "this approach denies poor people their capacity to organize and struggle for themselves. ... In NGO strategies, there has never been a sustained faith in the ability of poor people to bring about their own transformation. It is this perspective that has disallowed any real participation of the rural poor in NGO activities or the development of systems of accountability to them."

of autonomy and guard against cooptation, while at the same time, continue to be accountable to a number of stakeholders.

At the same time, NGOs must be aware of the pitfalls that accompany association with governments. As NGOs become more dependent on government and donor funding, they may find that their original agendas of development more closely reflect those of their funders than their rural constituents. Accountability can, as a result, turn into cooptation and corruption. This perception, on the part of a large numbers of NGOs, was one reason that the proposed Code of Conduct for Indian NGOs (see Chapter 4) was, and continues to be, so vehemently debated.

To provide some measure of accountability to outside partners while avoiding cooptation, NGOs must develop mechanisms for performance monitoring, accountability, and strategic planning such that their agreement to work with GOs does not result in their cooptation (unlike many NGOs, PRADAN has, since its creation, engaged in this kind of self-examination and as a result regularly modifies its operations accordingly). Such efforts are both time- and resource-consuming, and have been approached by only a small proportion of NGOs. Yet NGOs that fail to engage in such learning processes are in danger of losing a significant degree of their operational autonomy if and when they choose to work with their governments.

Institutionalizing NGO-GO Relations

If stakeholder accountability is of such importance to the creation of collaborations, it seems to follow that a mechanism by which members of both the NGO and government sectors can determine and monitor each others' suitability as partners should be made available.

As mentioned, a large proportion of discernible collaborations involve large, well-endowed NGOs which operate in extended geographic areas. Smaller NGOs with only limited areas of operations and resources are much less visible and influential to government policy-makers unless they can ally themselves with other, larger groups or successfully bring their concerns to the public arena through public protest. Many cooperative relationships—including critical collaborations—are established or facilitated through personal ties between GO and NGO officials, which means that the larger NGOs have an advantage in establishing such relationships. Arguably, personal relationships are the single most important factor in the successful creation and maintenance of collaborations; PRADAN finds such personal ties central to its relations with GOs at all levels, and notes that all its collaborative relationships are person-specific.

Although such relationships are viewed by many as an important factor in the creation of collaborations, others see the mechanism of particularistic person-to-person relationships as harmful to the NGO sector at large—that, by relying on such *ad hoc* linkages, governments are engaging in both patronage and discretional application of the law by practicing preferential treatment for certain NGOs. While this sort of treatment may be in an NGO's immediate interest, the cause of development—and the creation of a more accountable government—may be better served through the creation of more systematic, and less personalistic, forms of interaction.

If the creation of collaborations is dependent upon individualized personal contact, how is it possible to include the smaller, poorer, less vocal, and more geographically limited NGOs in the process of consideration for inclusion in collaborative efforts? While personal contacts may suffice for larger NGOs with the resources to include such efforts in their activities, there is less likelihood of creating relationships with the small organizations whose focus is on the smallest, poorest, least accessible villages and for whom collaboration could make a major difference in their efforts. How can a larger proportion of the NGO sector be identified and included in the selection process?

One answer would be to implement some standardized form of measuring NGOs' activities and their impacts, such as a code of conduct or an independent body to monitor NGO activities. The creation of a mechanism by which NGOs would report their activities and finances openly to government would "offer a responsible message to governments as to why governments must share this burden [of development] more broadly with all elements of society" (Riker 1995b: 124). Despite a general dislike of being required to make public details of their operations, NGOs would be well advised to become more open in their dealings with government if they wish to be seriously considered as partners in the policy process.

Another step to greater accountability would be to develop a set of indicators for assessing and monitoring the impact and outcomes of both GO and NGO interventions. While government operations are usually widely known, NGO activities are rarely well-publicized or subject to rigorous scrutiny. There is great deal of conjecture and word-of-mouth assertions about what they are able to achieve, but surprisingly little substantive review of NGOs' actual accomplishments. The concept of "social auditing"—adapting measures traditionally used to monitor financial practices for the purpose of scrutinizing the practices and accountability of rural development intervenors—is an innovative approach to such assessment (see Zadek and Gatward 1996).

It should be emphasized that these measures cannot replace the dynamic processes of collaboration. To suggest formalizing some of the processes that facilitate the creation of collaborative relationships does not mean that the collaborative process itself should be institutionalized, even if it were possible to do so. As the cases in the study show, the concept of critical collaboration is most accurately viewed as a process, not a single event; because the relationships between the stakeholders are dynamic, they are subject to renegotiation as situations change. The nature of the relationships between specific GOs and NGOs are highly context-specific, and attempts to routinize the content of the relationship would defeat the process itself, as well as the purpose for which it is intended.

The Future of Critical Collaboration

The measures proposed above are certainly not panaceas, nor is the accomplishment of such basic change easily or quickly done. The basis of such efforts, as well as the success of collaboration formation are all dependent on a sense of trust and a wish to work together among all stakeholders. GOs and NGOs may continue to work collaboratively despite an ongoing feeling of mutual antagonism. Also, one must not expect that once a collaborative relationship has been initiated that its success will automatically ensue. The processes that lead to mutually agreeable policies and practices among the stakeholders are subject to considerable and continual redefinition and negotiation—not only concerning the content and eventual outcome of a particular collaborative effort, but in delineating the nature of the problem itself and what the nature of the collaboration will comprise. The fluid nature of organizations and their interactions with each other, and other elements of the social, political and economic situations in which they work makes interrelationships necessarily mutable.

Given the diversity of the NGO sector, can a large number of individual collaborations between NGOs and government agencies at the various levels of government bring about significant advances in the accomplishment of rural development efforts? Can the experiences of isolated occurrences of collaboration be replicated and multiplied to the point that large-scale, even system-wide, change can take place? Fisher (1997) suggests that NGO networking at the grassroots, local, and national levels may create the "critical mass" necessary to challenge powerful interests within the government, although she despairs that even this will be successful in achieving such change. A partial answer to this concern, I believe, lies, in part, in the creation of alliances and networks of NGOs, through which numerous voices can be heard, and governments policies and activities can be challenged when necessary. Of equal importance, though, is the crafting of mutually agreeable formal arrangement in which individual, as well as allied, NGOs can approach governments in the spirit of cooperation and collaboration, as well as through the more advocative and confrontational approaches. Institutionalization of the selection process, in a manner that is mutually agreeable to both the GO and NGO sectors, is an important move in this direction.

In the final analysis, the possibility that the NGO and government sectors will actually work together appears to require at the very beginning the ability of both to overcome their fear and mistrust of each other. They must realize that neither governments, nor civil society, nor markets possess the knowledge or ability to accomplish on their own such a huge, multifaceted task as socio-economic development. Rather, each sector has something to offer. Only through a conscious decision to overcome their parochialism in favor of joining forces will it be possible for the various actors to begin to address in a new way not only the symptoms, but the causes, of the problems that continue to hinder the achievement of just livelihoods for their rural poor.

Surmounting their misgivings is, of course, not easily accomplished, nor are the learning processes that will allow both GOs and NGOs to engage in fruitful relationships. Yet movement toward greater NGO-GO interaction has been observed in a number of countries, and the limited availability of official funds and government personnel for implementing rural development interventions means that there will remain a role for NGOs to play.

Critical collaboration holds great promise. It is not a quick-fix to the shortcomings of current approaches to rural development, and efforts to make collaboration a wide-ranging part of the policy process will take a great deal of effort. However, if sincere efforts are made to create a truly responsive approach, critical collaboration may become more than yet another fleeting trend and, instead, contribute to real improvements in the lives of the rural poor of the Third World.

Bibliography

Adams, W.M. 1990. *Green Development: Environment and Sustainability in the Third World*. London and New York: Routledge.
Agarwal, Anil. 1985. "Politics of Environment-2," in *The State of India's Environment 1984–5: The Second Citizen's Report*. Delhi: Centre for Science and Environment.
———. 1986. "The Fifth World Conservation Lecture: Human-Nature Interactions in a Third World Country", *The Environmentalist*, 6(3): 165–183.
Ali, Sofi. 1992. "Public Policy-Making in India: Lessons from Rural Development Programmes", *Indian Journal of Public Administration*, 38(2): 109–121.
Alliband, Terry. 1983. *Catalysts of Development: Voluntary Agencies in India*. W. Hartford CT: Kumarian Press.
Almond, Gabriel A. and **Sidney Verba.** 1965. *The Civic Culture: Political Attitudes and Democracy in Five Nations, an Analytic Study*. Boston: Little, Brown and Co.
Altieri, Miguel A. 1990. "Agroecology and Rural Development in Latin America" in M. Altieri and Susanna B. Hecht, (eds.), *Agroecology and Small Farm Development*. pp. 113–120. Boston: CRC Press.
Alvares, Claude and **Ramesh Billorey.** 1987. "Damning the Narmada: The Politics Behind the Destruction," *The Ecologist*, 17(2): 62–72.
Ambler, John. 1994. "Financial and Institutional Challenges for Tank Sustainability: Some Historical Lessons," in *Discussion Papers: Seminar on Regeneration of Farmers' Management of Tank Irrigation Systems; March 1&2, 1994*. Ch. 9. Madurai, India: District Rural Development Agency and PRADAN.
Ananda, N. 2001. Personal communication.
APCG (Auroville Project Coordination Group). 1996. *Auroville and the Neighbouring Villages: A Presentation of Auroville's Activities in the Bioregion*. Auroville, India: Centre for Scientific Research.
Appa, Gautam. 1992. "Narmada Projects without World Bank Backing?", *Economic and Political Weekly*, 27(48): 2577–2580.
Appleby, Paul H. 1953. *Public Administration in India: Report of a Survey, 1953*. Delhi: Government of India Cabinet Secretariat, Manager of Publications.
Arendt, Hannah. 1958. *The Human Condition*. Chicago and London: The University of Chicago Press.

Arora, Balveer. 1987. "Party System and Federal Structures in India: Linkages and Issues," in Tarun Chandra Bose, (ed.), *Indian Federalism: Problems and Issues*, pp. 174–186. Calcutta: K.P. Bagchi & Company.

Ascher, William and Robert Healy. 1990. *Natural Resource Policy-making in Developing Countries: Environment, Economic Growth, and Income Distribution*. Durham NC and London: Duke University Press.

Auroville. 1996. "Introduction to Auroville," available from http://www.auroville-india.org/intro2av/intro2av.htm, accessed on 4 March 1999.

———. 1998. "Relationship Between Auroville and the Villages," available at http://www.auroville-india.org/outreach/villages/relations.htm, accessed on 4 March 1999.

Austin, Dennis. 1995. *Democracy and Violence in India and Sri Lanka*. London: The Royal Institute of International Affairs.

Austin, Granville. 1966. *The Indian Constitution*. New York: Oxford University Press.

Avina, Jeffrey. 1993. "The Evolutionary Life-cycles of Non-Governmental Development Organizations," *World Development*, 13(5): 453–474.

Bartelmus, Peter. 1994. *Environment, Growth and Development: The Concepts and Strategies of Sustainability*. London: Routledge.

Baxi, Upendra. 1986. "Two Threats to Democratic Social Action", *Mainstream*, 24(48): 25–32.

Bebbington, Anthony and John Farrington. 1993. "Governments, NGOs and Agricultural Development: Perspectives on Changing Inter-Organizational Relationships", *Journal of Development Studies*, 29(2): 199–219.

Bebbington, Anthony and Roger Riddell. 1997. "Heavy Hands, Hidden Hands, Holding Hands? Donors, Intermediary NGOs and Civil Society Organisations," in Hulme and Edwards, (eds.), *NGOs, States and Donors: Too Close for Comfort?* pp. 107–127. New York: St. Martin's Press in association with The Save the Children Fund.

Beckerman, Wilfred. 1992. "Economic Growth and the Environment: Whose Growth? Whose Environment?", *World Development*, 20(4) (April): 481–496.

Benjamin, Roger W. 1972. *Patterns of Political Development: Japan, India, Israel*. New York: David McKay Company Inc.

Berkes, Fikret. 1989. *Common Property Resources: Ecology and Community-Based Sustainable Development*. New York: Bethaven Press.

Bhatt, Anil. 1985. "Asian NGOs in Development: Their Role and Impact", in Noeleen Heyzer, James V. Riker and Antonio B. Quizon (eds), *Government-NGO Relations in Asia: Prospects and Challenges for People-Centred Development*, Ch. 4. Kuala Lumpur: Asian and Pacific Development Centre.

Bhat, K.V. and S. Satish. 1993. "NGO Links with the Karnataka State Watershed Development Cell: MYRADA and the PIDOW Project," in John Farrington and David J. Lewis (eds), *Non-Governmental Organizations and the State in Asia*, pp. 160–618. London: Routledge.

Bhattacharya, Dipankar. 1993. "New Challenges for Bengal Left: Panchayat Poll Pointers", *Economic and Political Weekly*, 17–24 July 1993: 1491–1494.

Bhushan, Bharat. 1998. "Pollution threat to Kodaikanal", *The Times of India*, Chennai, 13 April.

Biggs, Stephen D. and Arthur D. Neame. 1996. "Negotiating Room to Maneuver: Reflections Concerning NGO Autonomy and Accountability within the New

Policy Agenda", in Michael Edwards and David Hulme, (eds.), *Beyond the Magic Bullet: NGO Performance and Accountability in the Post-Cold War World*, pp. 40–52. West Hartford CT: Kumarian Press.

Bose, Nirmal. 1987. "A New Look for Centre-State Relations," in Tarun Chandra Bose, (ed.), *Indian Federalism*, pp. 34–43. Calcutta: K.P. Bagchi & Company.

Bose, Tarun Chandra. (ed.). 1987. *Indian Federalism: Problems and Issues*. Calcutta: K.P. Bagchi & Company.

Boulding, Kenneth E. 1989. *Three Faces of Power*. Newbury Park CA: Sage Publications.

Bowden, P. 1990. "NGOs in Asia: Issues in Development," *Public Administration and Development*, 10: 141–152.

Bowonder, B. 1983. "Management of Forest Resources in India," *Public Administration and Development*, vol. 3: 373–381.

Bragg, C. Kaye. 1996. "Influencing Indian Policy-Making for Women's Development: The NGO Sector's Role" (Initial draft prepared for delivery at the 1996 Annual Meeting of the Western Association of Asian Studies, October 1996).

Bratton, Michael. 1989a. "Beyond the State: Civil and Associational Life in Africa", *World Politics*, 41(3) (April): 407–430.

———. 1989b. "The Politics of Government-NGO Relations in Africa", *World Development*, 17(4): 569–587.

———. 1990. "Non-Governmental Organizations in Africa: Can They Influence Policy?" *Development and Change*, 21(1): 87–118.

Brodhead, T. and B. Herbert-Copley. 1988. *Bridges of Hope? Canadian Voluntary Agencies and the Third World*. Ottawa: North-South Institute.

Brown, L. David. 1991. "Bridging Organizations and Sustainable Development", *Human Relations*, 44(8): 807–831.

Brown, L. David and David C. Korten. 1989. *Understanding Voluntary Organizations: Guidelines for Donors*, World Bank Working paper WPS 258. Washington DC: World Bank.

Cameron, C. and J. Cocking. 1991. *The Evaluation of NGO Activities, Organizations, Methodology and Results*. London: ODA, Evaluation Department.

CAPART (Council for Advancement of People's Action and Rural Technology). *Action Plan to Bring About A Collaborative Relationship Between VOs and Government*. New Delhi: Council for Advancement of People's Action and Rural Technology.

———. 1999. *CAPART At a Glance*, available at http://www.nic.in/capart/glance2.htm, accessed on 16 April 1999.

Carroll, T. 1992. *Intermediary NGOs: The Supporting Link in Grassroots Development*. West Hartford CT: Kumarian Press.

Chamberlain, Heath B. 1993. "On the Search for Civil Society in China", *Modern China*, 19(2): 199–215.

Chambers, Robert. 1983. *Rural Development: Putting the Last First*. London: Longman Press.

———. 1994. "Participatory Rural Appraisal (PRA): Analysis of Experience", *World Development*, 22(9): 1253–1268.

Chandra, Asok. 1965. *Federalism in India*. London: George Allen and Unwin.

Chandrasekaran, C. 1997. "Alternative Approaches to Tank Rehabilitation", in *IMTI 1997. National Seminar on Farmers' Participation in Tank Rehabilitation and Management, January 23–24 1997*, Ch. 2.

Charlton, Roger and Roy, May. 1995. "NGOs, Politics and Probity; A Policy Implementation Perspective", *Third Word Quarterly*, 16(2): 237–255.

Charlton, Sue Ellen M. 1997. *Comparing Asian Politics: India, China and Japan*. Boulder CO: Westview Press.

Chaturvedi, Anil. 1988. *District Administration: The Dynamics of Discord*. New Delhi: Sage Publications.

Chhatre, Ashwini. 1994. "Draft Forest Regulation: Missing the Wood as well as the Trees", *Economic and Political Weekly*, 29(38): 2473–2474.

Clairmont, Frederic F. 1996. *The Rise and Fall of Economic Liberalism: The Making of the Economic Gulag*. Mapusa, Goa, India: The Other India Press, in association with the Third World Network.

Clark, John. 1991. *Democratizing Development: The Role of Voluntary Organizations*. London: Earthscan.

———. 1992. "Democratizing Development: NGOs and the State," *Development in Practice*, 2(3): 151–162.

———. 1993. "The State and the Voluntary Sector," *Human Resources Working Papers*, 12. Washington DC: The World Bank.

———. 1995. "The State, Popular Participation and the Voluntary Sector", *World Development*, 23(4): 593–601.

———. 1997. "The State, Popular Participation and the Voluntary Sector", in David Hulme and Michael Edwards, (eds.), *NGOs, States and Donors: Too Close for Comfort?*, pp. 43–58. New York: St. Martin's Press.

Cobb. R., J.K. Ross and M.H. Ross. 1976. "Agenda Building as a Comparative Political Process", *American Political Science Review*, 76(126): 127–128.

Colchester, Marcus. 1985. "The World Bank Ignores Human Suffering and is in Breach of International Law", *The Ecologist*, 15(5/6): 286–292.

Commins, Steven. 1997. "World Vision International and Donors: Too Close for Comfort?" in David Hulme and Michael Edwards, (eds.), *NGOs, States and Donors: Too Close for Comfort?*, pp. 140–155. New York: St. Martin's Press.

Conroy, Czech and Miles Litvinoff, (eds.). 1988. *The Greening of Aid: Sustainable Livelihoods in Practice*. London: Earthscan Publications Ltd.

Constantino-David, Karina. 1992. "The Philippine Experience in Scaling up", in Michael Edwards and David Hulme, (eds.), *Making a Difference: NGOs and Development in a Changing World*, pp. 137–147. London: Earthscan Publications Ltd.

Constitution of India. 1996. 15th ed. Lucknow: Eastern Book Company.

Conyers, D. and M. Kaul. 1990. "Strategic Issues in Development Management: Learning from Successful Experience. Part 1", *Public Administration and Development*, 10: 127–140.

Crocker, David A. 1991. "Toward Development Ethics", *World Development*, 19(5): 457–483.

Crombrugghe, Genevieve de, Howes, Mick, and Nieuwkirk, Marc. 1985. *An Evaluation of CEC Small Development Projects*. Brussels: le Collectif d'Exchanges pour la Technologie Appropriée (COTA).

Daly, Herman E. and John B. Cobb, Jr. 1989. *For the Common Good: Redirecting the Economy Toward Community, the Environment and a Sustainable Future*. Boston: Beacon Press.

Davies, Susanna and Melissa Leach. 1991. "Globalism versus Villagism: National and International Issues in Food Security and the Environment" in *IDS Bulletin*, 22(3): 43–50.

Day, R. and Day, J.V. 1977. "A Review of the Current State of Negotiated Order Theory: An Appreciation and a Critique", in *The Sociological Quarterly*, 18: 126–142.

de Araujo Castro, Joao Augusto. 1972. "Environment and Development: The Case of the Developing Countries", in David Kay and Eugene Skolnikoff, (eds.), *World Eco-Crisis: International Organizations in Response*, pp. 237–252. Madison WI: The University of Wisconsin Press.

de la Court, Thijs. 1990. *Beyond Brundtland: Green Development in the 1990s*. New York: New Horizons Press.

Desai, Bharat. 1998. "New States Divide but Rule" *India Today*, 23(28): 18–20.

de Santa Ana, Julio. 1994. "The Concept of Civil Society", *The Ecumenical Review*, 46(1): 3–11.

Devereux, Stephen and John Hoddinott, (eds.). 1992. *Fieldwork in Developing Countries*. Boulder CO: Lynne Rienner Publishers Inc.

Dhesi, Autar S. 1996. "From Centralised to Decentralised Development in India: The Communitarian Perspective", *Community Development Journal*, 31(3): 201–213.

Diaz-Albertini, Javier. 1990. *Development as Grassroots Empowerment: An Analytic Review of NGDO Programs in Lima, Peru*. New Haven CT: Program on Non-Profit Organizations, Institutions for Policy Studies, Yale University.

Dichter, Thomas W. 1986. "Demystifying 'Policy Dialogue: How Private Voluntary Organizations Can Have an Impact on Host Country Policies", *Findings '86*. Norwalk CT: Technoserve.

Dick, H.W. and P.J. Rimmer. 1980. "Beyond the Formal/Informal Sector Dichotomy: Towards an Integrated Alternative", *Pacific Viewpoint*, 21(1): 26–41.

Dikshit, Ramesh Dutta. 1975. "Post-1945 Asian Federations", *The Political Geography of Federalism*, Ch. 5. New York: John Wiley and Sons.

DRDA (District Rural Development Agency). 1999. *District Rural Development Agencies*, http://www.nic.im/rural/drda.htm, accessed on 3 September 1999.

Drèze, Jean and Amartya Sen. 1995. *India: Economic Development and Social Opportunity*. New Delhi: Oxford University Press.

———. (eds.). 1996. *Indian Development: Selected Regional Perspectives*. Oxford, UK: Oxford University Press.

Drèze, Jean and Haris Gazdar. 1996. "Uttar Pradesh: Burden of Inertia", in Jean Drèze and Amartya Sen, (eds.), *Indian Development: Selected Regional Perspectives*, pp. 33–128. Oxford, UK: Oxford University Press.

Duncan, Emma. 1995. "India Survey: 'Hello, world'," *The Economist*, 21 January: 3–4.

Dwivedi, O.P. 1977. "India: Pollution Control Policy and Programmes", in *International Review of Administrative Sciences*, 43: 128.

Dwivedi, O.P. and J. Nef. 1982. "Crises and Continuities in Development Theory and Administration: First and Third World Perspectives", *Public Administration and Development*, 2(1): 59–77.
Eade, D. 1993. Editorial, in *Development in Practice*, 3(3): 161–162.
Earle, Valerie. 1968. *Federalism: Infinite Variety in Theory and Practice*. Itasca, Illinois: F.E. Peacock.
Edwards, Michael and David Hulme, (eds.), 1992. *Making a Difference: NGOs and Development in a Changing World*. London: Earthscan Publications Ltd.
———. 1996a. "Too Close for Comfort? The Impact of Official Aid on Non-Governmental Organizations", *World Development*, 24(6) (June): 961–973.
———. (eds), 1996b., *Beyond the Magic Bullet: NGO Performance and Accountability in the Post-Cold War World*. West Hartford CT: Kumarian Press.
Elazar, Daniel J. 1993. "International and Comparative Federalism", in *Political Science and Politics*, 26(2): 190–196.
Eldersveld, Samuel J., Vedula Jagannadham and A.P. Barnabas. 1968. *The Citizen and the Administrator in a Developing Democracy*. Chicago: Scott, Foresman.
Emery, F.E. and E.C. Trist. 1965. "The Causal Texture of Organizational Environments", *Human Relations*, 18(1): 125–34.
Erskine, A. 1999. Personal communication.
Escobar, Arturo. 1995. *Encountering Development: The Making and Unmaking of the Third World*. Princeton NJ: Princeton University Press.
Esman, Milton J. 1991. *Management Dimensions of Development: Perspectives and Strategies*. West Hartford CT: Kumarian Press.
Esteva, Gustavo and Madhu Suri Prakash. 1992. "Grassroots Resistance to Sustainable Development: Lessons from the Banks of the Narmada", *The Ecologist*, 22(2): 45–51.
Falk, Richard. 1995. *On Humane Governance, Toward a New Global Politics: The World Order Models Project Report of the Global Civilization Initiative*. University Park PA: Pennsylvania State University Press.
Farrington, John and Anthony Bebbington with Kate Wellard and David J. Lewis. 1993. *Reluctant Partners? Non-Governmental Organizations, the State and Sustainable Agricultural Development*. London: Routledge.
Farrington, John and S. Biggs. 1990. "NGOs, Agricultural Technology and the Rural Poor", *Food Policy*, (December): 479–492.
Farrington, John and David J. Lewis, (eds.). 1993. *Non-Governmental Organizations and the State in Asia: Rethinking Roles in Sustainable Agricultural Development*. London: Routledge.
Fernandez, Aloysius. 1987. "NGOs in South Asia", *World Development*, 15 (supplement): 39–49.
Fernandez, Aloysius and James Mascarenhas. 1993. "Mysore Relief and Development Agency (MYRADA): Participatory Rural Appraisal and Participatory Learning Methods", in John Farrington and David J. Lewis, *Non-Governmental Organizations and the State in Asia*, pp. 160–168. London: Routledge.
Fisher, Julie. 1997. *Non-Governments: NGOs and the Political Development of the Third World*. West Hartford, CT: Kumarian Press.
Fowler, Alan. 1991. "The Role of NGOs in Changing State-Society Relations: Perspectives from Eastern and Southern Africa", *Development Policy Review*, 9: 53–84.

Fox, J. 1992. "Democratic Rural Development: Leadership Accountability in Regional Peasant Organization", *Development and Change*, 23(2): 1–36.
Frankel, Francine. 1971. *India's Green Revolution: Economic Gains and Political Costs*. Princeton NJ: Princeton University Press.
Frantz, Telmo Rudi. 1987. "The Role of NGOs in the Strengthening of Civil Society", *World Development*, 15 (Supplement): 121–127.
Freire, Paulo. 1987. *Pedagogy of the Oppressed*. New York: Continuum.
———. 1989. *Learning to Question: A Pedagogy of Liberation*, trans. Tony Coates. New York: Continuum.
Friedrich, Carl J. 1986. *Trends of Federalism in Theory and Practice*. New York: Praeger Publishers.
Gablier, Thomas. 1997. Personal communication (13, 25, 26 February).
———. 1998. Personal correspondence.
———. 1999. Personal correspondence.
Gadgil, Madhav and Ramachandra Guha. 1994. "Ecological Conflicts and the Environmental Movement in India", *Development and Change*, 25: 101–136.
———. 1995. *Ecology and Equity: The Use and Abuse of Nature in Contemporary India*. New Delhi: Penguin Books India (P) Ltd.
Garilao, Ernesto D. 1987. "Indigenous NGOs as Strategic Institutions: Managing the Relationship with Government and Resource Agencies", *World Development*, 15 (supplement): 113–120.
Giordano, Edward. 1997. Personal communication.
Giordano, Edward, S. Satish and John Farrington. 1993. "Greenwork at Auroville: From Survival to Inter-Institutional Collaboration", in Farrington, John and David J. Lewis, (eds.), *Non-Governmental Organizations and the State in Asia: Rethinking Roles in Sustainable Development*, pp. 145–151. London, New York: Routledge.
Gorman, Robert F. 1984. *Private Voluntary Organizations as Agents of Development*. Boulder CO and London: Westview Press.
Government of India. 1988. *The Auroville Foundation Act*. Delhi: Ministry of Law and Justice.
Gray, Barbara. 1985. "Conditions Facilitating Interorganizational Collaboration", *Human Relations*, 38(10): 911–936.
———. 1989. *Collaborating: Finding Common Ground for Multiparty Problems*. San Francisco: Jossey-Bass Inc Publishers.
Gray, Barbara and Donna J. Wood. 1991. "Collaborative Alliances: Moving from Practice to Theory", Introduction, *The Journal of Applied Behavioral Science*, 27(1): 3–22.
Grove, Richard H. 1995. *Green Imperialism: Colonial Expansion, Tropical Island Edens and the Origins of Environmentalism*. Cambridge: Cambridge University Press.
Guimarães, Roberto P. 1991. *Ecopolitics of Development in the Third World: Politics and Environment in Brazil*. Boulder: Lynne Rienner Publishers, Inc.
Gupta, Dipankar. 1997. "Civil Society in the Indian Context: Letting the State off the Hook", *Contemporary Sociology*, 26(3): 305–307.
Gupta, Ranjit. 1986. "Similar Concern but Varying Approaches: A Case Study of ASSEFA and MYRADA", in Dantwala, M.L. et al. (eds.), *Asian Seminar on*

Rural Development: The Indian Experience, pp. 311–337. New Delhi: Oxford XIBH Publishing Company Pvt. Ltd.

Guthrie, Robin. 1994. *Civic, Civil or Servile?* Geneva: Interphil.

Hadenuis, Axel and Frederik Uggla. 1996. "Making Civil Society Work, Promoting Democratic Development: What Can States and Donors Do?" *World Development*, 24(10): 1621–1639.

Hardgrave, Robert L, Jr. 1987. "Social Change and Political Stability in India", Ch. 1 in Sankarankutty Nair, (ed.), *Modern India: Society and Politics in Transition*, pp. 49–66. New Delhi: Inter-India Publications.

Hardgrave, Robert L, Jr. and Stanley Kochanek. 1993. *India: Government and Politics in a Developing Nation*, 5th ed. New York: Harcourt Brace Jovanovich Inc.

Hashcmi, Syed M. 1996. "NGO Accountability in Bangladesh: Beneficiaries, Donors, and the State", in Edwards and Hulme, (eds), *Beyond the Magic Bullet: NGO Performance and Accountability in the Post-Cold War World*. Hartford CT: Kumarian Press.

Hayter, Theresa. 1981. *The Creation of World Poverty*. London: Pluto Press.

Healy, John M. and Mark Robinson. 1994. *Democracy, Governance and Economic Policy: Sub-Saharan Africa in Comparative Perspective*. London: Overseas Development Institute.

Heginbotham, Stanley J. 1975. *Cultures in Conflict: The Four Faces of Indian Bureaucracy*. New York: Columbia University Press.

Heyzer, Noeleen. 1995. "Toward New Government-NGO Relations for Sustainable and People-Centred Development", in Noeleen Heyzer et al., (eds.), *Government-NGO Relations in Asia: Prospects and Challenges for People-Centred Development*, pp. 1–13. Houndsmill, Basingstroke, Hampshire: MacMillan Press Ltd.

Hindu, The. 1988. "Holiday complex restrained from letting effluents into Kodai Lake", *The Hindu*, 15 April.

———. 1994. "Secret Sanction for Unauthorized Building?" *The Hindu*, Wednesday 22 June.

Howarth, David. 1995. "Discourse Theory", in David Marsh and Gerry Stoker, (eds.), *Theory and Methods in Political Science*, pp. 115–133. New York: St. Martin's Press.

Hulme, David and Michael Edwards, (eds.), 1997. *NGOs, States and Donors: Too Close for Comfort?* New York: St. Martin's Press, Scholarly and Reference Division.

Huntington, Samuel. 1968. *Political Order in Changing Societies*. New Haven: Yale University Press.

Hutchful, Eboe. 1996. "The Civil Society Debate in China", *International Journal*, 51(1): 54–77.

IMTI (Irrigation Management Training Institute). 1997. National Seminar on Farmers' Participation in Tank Rehabilitation and Management, 23–24 January 1997.

Jain, L.C. 1985. *Grass Without Roots: Rural Development Under Government Auspices*. New Delhi and Beverly Hills CA: Sage Publications.

Jain, R.B. and O.P. Dwivedi. 1990. "Policy Developments and Administrative Change in India", in O.P. Dwivedi and Keith M. Henderson, *Public Administration in World Perspective*, pp. 166–198. Ames IA: Iowa State University Press.

Jaix, Randhir B. 1992. "Role of Residents, Non-Governmental Organizations and Quasi-Public Agencies in Local Government in India", in *Comparative Studies of Public Administration: The Role of Residents, Non-Governmental Organizations and Quasi-Public Agencies*, pp. 88–113. Tokyo: Eropa Local Government Center.

Jayal, Niraja Gopal. 1997. The Government Agenda: Making Democratic Development Dispensable", *Economic and Political Weekly*, 32(8) (February 22): 407–412.

Jeyakaran, C. 1996, 1997. Personal communication.

Judge, Anthony J.N. 1995. "NGOs and Civil Society: Some Realities and Distortions—the Challenge of 'Necessary-to-Govern Organizations'", *Transnational Associations*, 47 (May–June): 156–180.

Kalimullah, Nazmul Ahsan. 1989. "The Political Environment of Non-Governmental Organisations in India: A Study of Three States," in Chaudhuri, Ranjit, (ed.), *Conflict, Development and NGOs*, pp. 95–103. Calcutta: Best Books.

Kannan, K.P. 1993. "Panchayat Raj Act: Local Self-Government and Decentralised Development", *Economic and Political Weekly*, December 4: 2644–2646.

Kannan, R. 1998. Personal communication.

Kanavalli, S. 1993. *Quest for Justice: An account of a people's struggle for reclaiming their village common lands taken over by a joint-sector company of an industrial giant and a state government*. Dharwad, India: Samaj Parivartana Samudaya.

Keane, John. 1988. *Democracy and Civil Society*. London: Verso Press.

Khan, Rasheeduddin. 1997. "Federalism in India: A Quest for New Identity", in Rasheeduddin Khan, (ed.), *Rethinking Indian Federalism*. Shimla, India: Inter-University Centre for Humanities and Social Sciences.

Khator, Renu. 1984. "Environment as a Political Issue in Developing Countries: A Study of Environmental Pollution in India—A Viewpoint," *International Journal of Environmental Studies*, 23: 105–112.

———. 1991. *Environment, Development and Politics in India*. Lanham MD: University Press of America.

Khilnani, Sunil. 1997. *The Idea of India*. New York: Farrar, Straus and Giroux.

Kiggundu, Moses N. 1989. *Managing Organizations in Developing Countries: An Operational and Strategic Approach*. West Hartford CT: Kumarian Press.

Kohli, Atul. 1984. "Communist Reformers in West Bengal: Origins, Features, and Relations with New Delhi", in John R. Wood, (ed.), *State Politics in Contemporary India: Crisis or Continuity?* pp. 93–96. Boulder CO: Westview Press.

Korten, David C. 1986. "Micro-Policy Reform: The Role of Private Voluntary Development Agencies", in *NASPAA Working Paper No. 12*. Washington DC: National Association of Schools of Public Affairs and Administration.

———. 1987. "Third Generation NGO Strategies: A Key to People-Centered Development", *World Development*, 15 (Supplement): 145–159.

———. 1990. *Getting to the 21st Century: Voluntary Action and the Global Agenda*. West Hartford CT: Kumarian Press.

Korten, David and **Antonio B. Quizon.** 1995. "Government, NGO and International Agency Cooperation: Whose Agenda?" in Heyzer et al., (eds.), *Government-NGO Relations in Asia: Prospects and Challenges for People-Centred Development.* Kuala Lumpur: Asian and Pacific Development Centre.

Kothari, Rajni. 1961. "Administrative Institutions of Government", *The Economic Weekly,* May 27: 819–826.

———. 1988. *State against Democracy: In Search of Humane Governance.* Delhi: Ajanta Publications (India).

———. 1989. *Rethinking Development: In Search of Humane Alternatives.* New York: New Horizons Press.

Kothari, Smithu. 1993. "Social Movements and the Redefinition of Democracy", in Philip Oldenburg, (ed.), *India Briefing, 1993,* pp. 131–162. Boulder CO: Westview Press.

Kottack, C.P. 1985. "When People Don't Come First: Some Sociological Perspectives from Completed Projects," in Cernea, Michael M., (ed.), *Putting People First: Sociological Variables in Rural Development.* New York: Oxford University Press for the World Bank.

Krishna, Sumi. 1996. *Environmental Politics: People's Lives and Development Choices.* New Delhi: Sage Publications.

Kumar, Girish. 1995. "Postponement of Tamil Nadu's Panchayat Election: Centre's Role", *Economic and Political Weekly,* 9 September: 2235–2236.

Lacey, Hugh. 1991. "Human Rights and Authentic Development: A Response to Nickel", in Kenneth Aman, (ed.), *Ethical Principles for Development: Needs, Capacities or Rights?* pp. 213–218. Upper Montclair NJ: The Institute for Critical Thinking.

Lawrence, P.R. and **J.W. Lorsch.** 1967. *Organization and Environment.* Cambridge MA: Harvard University Press.

Leach, Melissa. 1988. "Becoming a People Sector: Indian Voluntary Agencies at a Crossroads", *Community Development Journal,* 23(2): 86–93.

Lele, Sharachchandra. 2000. "Godsend, Sleight of Hand, or Just Muddling Through: Joint Water and Forest Management in India", Overseas Development Institute Natural Resources Perspectives no. 53, available at http://www.oneworld.org/odi/nrp/53.html, accessed on 18 December 2000.

Lewis, John P. 1962. *Quiet Crisis in India: Economic Development and American Policy.* Washington DC: Brookings Institution.

Lewis, W. Arthur. 1963. "Economic Development with Unlimited Supplies of Labour", in A.N. Agarwala and S.P. Singh, (eds.), *The Economics of Underdevelopment.* New York: Oxford University Press.

Lijphart, Arend. 1971. "Comparative Politics and the Comparative Method", *American Political Science Review,* 65(3): 682–693.

Livernash, Robert. 1992. "The Growing Influence of NGOs in the Developing World", *Environment,* 34(5): 13–20, 41–43.

Logsdon, Jeanne M. 1991. "Interests and Interdependence in the Formation of Social Problem-Solving Collaborations", *The Journal of Applied Behavioral Science,* 27(1): 23–37.

Maheshwari, Shiriam. 1990. "India", in V. Subramanian, (ed.), *Public Administration in the Third World: An International Handbook,* Ch. 2. Westport CT: Greenwood Press.

Maniates, Michael. 1990. *Organizing for rural energy development: improved cookstoves, local organizations, and the state in Gujarat, India*. Berkeley CA: University of California, Berkeley.
———. 1992. "State-NGO collaboration in rural resource management: India's improved cookstove programme," *Unasylva*, 43: 21–29; also available at http:www.fao.org/docrep/u7760e05.htm#state accessed on 7 January 2001.
Manor, James. 1990. "How and why liberal and representative politics emerged in India", *Political Studies*, 38: 20–38.
Mars, Theodor Friedrich. 1992. "Public Sector Organization: Where Next?" *IDS Bulletin*, 23(4) (October 1): 18–30.
Marsden, David and **Peter Oakley.** 1990. *Evaluating Social Development Projects*, Development Guidelines No. 5. Oxford: Oxfam.
Mathiot, Elizabeth Moen. 1998. *NGOs and Grassroots Development Work in South India*. Lanham MD: University Press of America Inc.
McCarthy, Jane E with **A. Shorett.** 1984. *Negotiating Settlements: A Guide to Environmental Mediation*. New York: American Arbitration Association.
Meinzen-Dick, Ruth Suseela. 1984. *Local Management of Tank Irrigation in South India: Organization and Operation*, Cornell Studies in Irrigation, No. 3. Ithaca NY: Cornell University.
Mencher, John. 1999. "NGOs: Are They a Force for Change?" *Economic and Political Weekly*, 34(30): 2081–2086.
Minor, Robert N. 1999. *The Religious, the Spiritual and the Secular: Auroville and Secular India*. Albany: State University of New York.
Mitra, Subrata K. 1992. *Power, Protest and Participation: Local Elites and the Politics of Development in India*. London: Routledge.
Moen, Elizabeth. 1991. *Voluntary Sector Grass Roots Development in Tamil Nadu*. Tamil Nadu, India: Gandhigram Rural Institute.
Mukhia, H. 1989. "Dynamics of Rural Power," *Economic and Political Weekly*, (17 June): 1338.
Mukhopadhyay, Asok Kumar. 1987. "Federalism and All-India Services", in Tarun Chandra Bose, (ed.), *Indian Federalism: Problems and Issues*. Calcutta: K.P. Bagchi & Co.
NAEB (National Afforestation and Eco-Development Board). 1996. *Guidelines for Grants-in-Aid Scheme*. New Delhi: Ministry of Environment and Forests.
Nagle, David. 1997. Personal communication.
Najam, Adil. 1996. *Nongovernmental Organizations as Policy Entrepreneurs*. New Haven CT: Program on Non-Profit Organizations, Institution for Social and Policy Studies, Yale University.
Nandy, Ashis. 1989. "The Political Culture of the Indian State", *Daedalus*, 118 (Fall): 1–26.
Ndegwa, Stephen N. 1996. *The Two Faces of Civil Society: NGOs and Politics in Africa*. W. Hartford CT: Kumarian Press.
Nef, Jorge and **O.P. Dwivedi.** 1981. "Development Theory and Administration: A Fence Around an Empty Lot?", *Indian Journal of Public Administration*, 27(1): 42–66.
Nehru, Jawaharlal. 1960. *A Bunch of Old Letters*. Bombay: Asia Publishing House.

Nemeth, C. 1970. "Bargaining and Reciprocity", *Psychological Bulletin,* 74: 297–308.
Nerfin, Marc. 1986. "Neither Prince nor Merchant—Citizen: An Introduction to the Third System", *IFDA Dossier,* 56 (November/December): 3–28.
Noman, Omar. 1996. *Economic Development and Environmental Policy.* London; New York: Kegan Paul International.
Pai, M.P. 1980. *The Emerging Role of the Collector.* Delhi: Indian Institute of Public Administration.
Paul, Samuel. 1988. "Governments and Grassroots Organizations: From Coexistence to Collaboration", in J.P. Lewis, (ed.), *Strengthening the Poor: What Have We Learned?* pp. 61–72. New Brunswick, NJ: Transaction Books, Overseas Development Council.
———. 1989. "Poverty Alleviation and Participation: The Case for Government-Grassroots Agency Collaboration", *Economic and Political Weekly,* 24(2): 100–105.
Pearce, Jenny. 1996. "Between Co-option and Irrelevance? Latin American NGOs in the 1990s", in David Hulme and Michael Edwards, (eds.), *NGOs, States and Donors: Too Close for Comfort?* pp. 257–274. New York: St. Martin's Press, in association with Save the Children.
Penz, Peter. 1986. *Consumer Sovereignty and Human Interests.* Cambridge UK: Cambridge University Press.
Pieterse, Jan Nederveen. 1991. "Dilemmas of Development Discourse: The Crisis of Developmentalism and the Comparative Method", *Development and Change,* 22(1): 5–29.
Potter, David. 1995. *NGOs and Forest Management in Karnataka,* GECOU Working Paper No. 3. Milton Keynes, UK: The Open University.
———. 1996a. *India's Political Administrators: From ICS to IAS.* New Delhi: Oxford University Press.
———. 1996b. "Democratisation and the Environment: NGOs and Deforestation Policies in India (Karnataka) and Indonesia (North Sumatra)", in David Potter, (ed.), *NGOs and Environmental Policies: Asia and Africa,* pp. 9–37. Portland OR: Frank Cass.
Powell, Walter W. and **Paul J. DiMaggio,** (eds.), 1991. *The New Institutionalization in Organizational Analysis.* Chicago and London: The University of Chicago Press.
PRADAN. 1986. *Uttarkhand Hill Region: Problems and Prospects,* in association with the Tibetan Refugee Aid Society. Unpublished.
———. 1988. *On Enhancing Capabilities of Voluntary Agencies.* A report submitted to the Council for the Advancement of People's Action and Rural Technology (CAPART), New Delhi.
———. 1990. *In Search of Significance–Ten Years of Human Resource Development for the Rural Sector,* in association with The Institute of Rural Management, Anand. Unpublished.
———. 1991. *Pilot Project for Improving the Effectiveness of Poverty Alleviation Programmes at the Block Level, Final Report.* Unpublished.
———. 1995. *Annual Report, Balance Sheet and Income and Expenditure Account for the period ended 31 March 1995.* Unpublished.
———. 1996a. *Case Study of PRADAN.* New Delhi: Unpublished manuscript.

PRADAN. 1996b. *Towards A Relationship of Significance: Interim Report on the Study of Relationships Between Government and NGOs in Rajasthan.* New Delhi: Professional Assistance for Development Action.

———. 1996c. *Search for Institutional Mechanisms to Build a Sustainable Relationship between Government of Rajasthan and the Private Voluntary Initiatives for Social Development in the State: A Proposal to the Ford Foundation.* Unpublished.

———. 1997. Confidential communication.

Ragin, Charles C. 1987. *The Comparative Method: Moving Beyond Qualitative and Quantitative Strategies.* Berkeley and Los Angeles: University of California Press.

Rahman, Anisur, (ed.), 1984. *Grass Roots Participation and Self-Reliance: Experiences in South and South East Asia.* New Delhi: Oxford and IBH Publishing Co.

Rajasekhar, D. 1997. Personal communication.

Rajasekhar, D. and Lini R. Nair. 1995. "Networks among NGDOs in Karnataka", in D. Rajasekhar, (ed.), *People Centered Development: NGDOs and Decentralised Government,* Ch. 6. Bangalore, India: Institute for Social and Economic Change.

Rajivan, Krishnaswamy. 1997. Personal communication.

Ramaswamy, S.R. and G.N.N. Prasad. 1990. *An Experiment in Working Together.* Bangalore: FEVORD-K.

Rao, J. Mohan. 1995. "Whither India's Environment?", *Economic and Political Weekly,* 30(13): 677–686.

Ray, Amal. 1966. *Inter-Governmental Relations in India.* New York: Asia Publishing House.

Ray, Bharati. 1987. "Centre-State Financial Relations in India: As Visualised by the Framers of the Constitution and as in Practice", in Tarun Chandra Bose, (ed.), *Indian Federalism:* 111–124.

Redclift, Michael R. 1987. *Sustainable Development: Exploring the Contradictions.* London; New York: Methuen.

Reddy, N.L. Narasimha and D. Rajasekhar. 1996. *Development Programmes and NGOs: A Guide on Central Government Programmes for NGOs in India.* Bangalore, India: Bangalore Consultancy office, and NOVIB, The Hague, The Netherlands.

Rengasamy, S. 1997. Personal communication.

Riddell, Roger C. and Mark Robinson. 1995. *Non-Governmental Organizations and Rural Poverty Alleviation.* Oxford: Clarendon Press.

Riker, James V. 1995a. "Contending Perspectives for Interpreting Government-NGO Relations in South and Southeast Asia: Constraints, Challenges and the Search for Common Ground in Rural Development", in Noeleen Heyzer et al., *Government-NGO Relations in Asia: Prospects and Challenges for People-Centred Development,* pp. 15–55. Kuala Lumpur: Asian and Pacific Development Centre.

———. 1995b. "From Cooptation to Cooperation and Collaboration in Government-NGO Relations: Toward an Enabling Policy Environment for People-Centred Development in Asia", in Noeleen Heyzer et al., *Government-NGO Relations in Asia: Prospects and Challenges for People-Centred Development,* pp. 91–130. Kuala Lumpur: Asian and Pacific Development Centre.

Robinson, Mark A. 1991. *Evaluating the Impact of NGOs in Rural Poverty Alleviation: India Country Study,* ODI Working Paper No. 49. London: Overseas Development Institute.

Robinson, Mark A. 1992. "Assessing the Impact of NGO Rural Poverty Alleviation Programmes: Evidence from South India", *Journal of International Development*, 4(4): 397–417.

Robinson, Mark, John Farrington and S. Satish. 1993. "Chapter 4, Overview", in Farrington, John and David J. Lewis, (eds.), *Non-Governmental Organizations and the State in Asia: Rethinking Roles in Sustainable Agricultural Development*, pp. 91–101. London: Routledge.

Rondinelli, Dennis A. *Development Projects as Policy Experiments: An Adaptive Approach to Development Administration*, 2nd ed. London: Routledge.

Rostow, W.W. 1960. *The Stages of Economic Growth: A Non-Communist Manifesto*. London: Longman Inc.

Roy, Bunker. 1993. "NGOs Too Need a Code", *The Indian Express*, 29 December: 8–9.

———. 1995. "Threats to Volunteerism: Why a Code of Conduct is Necessary", *Indian Journal of Social Science*, 56(1): 35–37.

———. 1996. "Open Letter to Home Minister: Foreign Funds and Threat to Voluntary Sector", *Economic and Political Weekly*, 7 December: 3161–3162.

———. 1998. "Vanishing Voluntarism", *Himal: The South Asian Magazine*, 11(8): 44–45.

Rudolph, Lloyd I. and Susanne Hoeber Rudolph. 1987. *In Pursuit of Lakshmi: The Political Economy of the Indian State*, Chicago: The University of Chicago Press.

Sachs, Wolfgang. 1988. "The Gospel of Global Efficiency: On Worldwatch and Other Reports on the State of the World", *IFDA Dossier*, 68 (November/December): 33–41.

———. 1990. "The Archaeology of Development Idea", *IFDA Dossier*, No. 68: 33–39.

———. 1993. *Global Ecology: A New Arena of Political Conflict*. London: Zed Books.

Sadasivan, S.N. 1980. *Politics and Administration in India*. Delhi: Indian Institute of Public Administration.

Sainath, P. 1996. *Everybody Loves a Good Drought: Stories from India's Poorest Districts*. New Delhi: Penguin Books India (P) Ltd.

Saint-Martin, Jean-Guy. 1994. "Increasing NGO Influence on Development Policy", *Peace Research*, 26(1): 79–86.

Saktivadivel, R., Shanmugham, C.R. and Savadamuthu, S. 1982. "A Pilot Project Study of the Modernization of Tank Irrigation in Tamil Nadu", in *Proceedings of the International Workshop on Modernization of Tank Irrigation: Problems and Issues*, pp. 1–21. Madras: Centre for Water Resources, College of Engineering, Perarignar Anna University of Technology.

Saltyagopal, K. 1997. Former District Collector, Madurai District. Personal communication.

Sanpathy, T. 1984. *Irrigation and Economic Development*, New Delhi: Ashish Publishing House.

Sanyal, Bishwapriya. 1991. "Antagonistic Cooperation: A Case Study of Non-governmental Organizations, Government and Donors' Relationships in Income-Generating Projects in Bangladesh", *World Development*, 19(10): 1367–1379.

Sastry, K.R. 1995. "Panchayat Raj Act: Without Substance", *Economic and Political Weekly*, 30(30): 1909–1910.

Schaffer, B.B. 1969. "The Deadlock in Development Administration", in Colin Leys, (ed.), *Politics and Change in Developing Countries*, pp. 177–211. Cambridge: Cambridge University Press.

Schramm, W. 1971. *Notes on Case Studies of Instructional Media Projects*, Working Paper, The Academy for Educational Development, Washington D.C.

Schumacher, E.F. 1973. *Small is Beautiful: Economics as if People Really Mattered.* London: Blond and Briggs.

Segal, Jerome M. 1991. "The Economic Development Problem", in Kenneth Aman, (ed.), *Ethical Principles for Development: Needs, Capacities or Rights?* pp. 213–218. Upper Montclair NJ: The Institute for Critical Thinking.

Sen, Amartya. 1984. *Resources, Values and Development.* Cambridge: Harvard University Press.

———. 1987. "The Standard of Living II, Lives and Capabilities", in Hawthorne, Jeffrey, (ed.), *The Standard of Living.* Cambridge: Cambridge University Press.

———. 1988. "The Concept of Development", in Hollis Chenery and T.N. Srinivasan, (eds.), *Handbook of Development Economics*, vol. 1, pp. 9–26. Amsterdam: North Holland Publishers.

Sen, Biswajit. 1987. "NGO Self-evaluation: Issues of Concern", *World Development*, 15 (supplement): 161–167.

Sen, S.R. 1988. "Centre-State Relations in India", *Economic and Political Weekly*, 6 August: 1637–1641.

Sengupta, Prasanta. 1987. "Inter-Governmental Relationship: Crisis in 'Federal Spirit'", in Tarun Bose, ed., *Indian Federalism*, 137–147.

Sengupta, Sunil and Haris Gazdar. 1996. "Agrarian Politics and Rural Development in West Bengal", in Jean Drèze and Amartya Sen, (eds.), *Indian Development: Selected Regional Perspectives*, pp. 129–204. Oxford, UK: Oxford University Press.

Shams, Rasul. 1994. "Environmental Policy and Interest Groups in Developing Countries", *Intereconomics*, (January/February): 16–24.

Shanmugam, C.R. 1996. Personal communication.

Shaw, Bernard. 1990. *Man and Superman*. Norwalk CT: Heritage Press.

Sheth, D.L. 1987. "Alternative Development as Political Practice", *Alternatives*, 12(2): 155–171.

Sheth, D.L. and Harsh Sethi. 1991. "The NGO Sector in India: Historical Context and Current Discourse", *Voluntas*, 2(2): 49–68.

SHIELD (Sempatti Hill Initiators for Ecological Development). 1996. *Annual Report for 1995–96.* Unpublished.

Shourie, H.D. 1995. "Too Many Cooks Spoil the Broth", *The Sunday Times of India*, (18 June): 6.

Shue, Henry. 1980. *Basic Rights: Subsistence, Affluence and U.S. Foreign Policy.* Princeton NJ: Princeton University Press.

Siffin, William J. 1976. "Two Decades of Public Administration in Developing Countries", *Public Administration Review*, 36(1): 61–71.

Singh, Hoshiar. 1994. "Constitutional Base for Panchayati Raj in India: The 73rd Amendment Act", *Asian Survey*, 34(9): 818–827.

———. 1995. *Administration of Rural Development in India.* New Delhi: Sterling Publishers Private Ltd.

Smith, W.E., F.J. Lethem, and B.A. Thoolen. 1981. "The Design of Organizations for Rural Development Projects A Progress Report", *World Bank Staff Working Paper*, No. 375.

Sollis, P. 1991. "Multilateral Agencies and NGOs in the Context of Policy Reform", Paper presented at the Conference on *Changing US and Multilateral Policy Toward Central America*, 10–12 June 1992, Washington DC.

SPEECH (Society for Peoples Education and Economic Change). *Statement of Purpose.*

Stone, Donald C. 1965. "Government Machinery Necessary for Development", in M. Kriesberg, (ed.), *Public Administration in Developing Countries*, pp. 49–67. Washington DC: The Brookings Institution.

Subramanian, G. Ganapathi. 1997. "An Overview of Tank System Management", in Selvaraj, K.V. and P. Sankarakumar, (coordinators), *National Seminar on Farmers' Participation in Tank Rehabilitation and Management*, 24–25 January 1997. Thiruchirapalli, India: Irrigation Management Training Institute.

Tamil Nadu Forest Department. 1995. *Management Agreement Between the Forest Department and the Palani Hills Conservation Council for Cooperation in the Implementation of the Kadavakurichi Interface Forestry Project*. Unpublished.

Tandon, Rajesh. 1987. "The Relationship Between NGOs and the Government", quoted in Clark 1992. "Democratising Development: NGOs and the State", *Development in Practice*, 2(3) (October): 151–162.

———. 1991. *NGO-Government Relations: A Source of Life or a Kiss of Death?* Delhi: Society for Participatory Research in India.

———. 1993. "NGO Regulation: South Asian Scenario", *Transnational Associations*, 4: 200–206.

———. 1996. "Board Games: Governance and Accountability in NGOs", in Michael Edwards, and David Hulme, (eds.), *Beyond the Magic Bullet: NGO Performance and Accountability in the Post-Cold War World*, pp. 53–63. West Hartford CT: Kumarian Press.

Tendler, Judith. 1982. *Turning Private Voluntary Organizations into Development Organizations: Questions for Evaluation*, AID Program Evaluation Discussion Paper No. 12. Washington DC: US Agency for International Development.

———. 1987. "Whatever Happened to Poverty Alleviation?", Report for the Mid-Decade Review of the Ford Foundation's Program on Livelihood, Employment and Income Generation, reprinted in *World Development*, 17(1): 1033–1044.

Thakkar, Usha and Mangesh Kulkarni. 1991. "Environment and Development: The Case of the Sardar Sarovar Project", *South Asia Bulletin*, 14(2): 96–103.

Thakur, Ramesh. 1995. *The Government and Politics of India*. New York: St. Martin's Press.

Thurow, Lester. 1980. *The Zero Sum Society: Distribution and the Possibilities for Economic Change*. New York: Basic Books.

Todaro, Michael P. 1985. *Economic Development in the Third World*, 3rd ed. New York: Longman Inc.

Trist, E.L. "Referent Organizations and the Development of Interorganizational Domains", *Human Relations*, 36(3): 247–268.

Tummala, Krishna K. 1994. *Public Administration in India*. Singapore: Times Academic Press.

Turner, Mark and David Hulme. 1997. *Governance, Administration and Development: Making the State Work*. West Hartford CT: Kumarian Press.
UNESCO. 1984. *Records of the General Conference: Twenty-second session*. Paris: UNESCO.
Unia, Pramod. 1991. "Social Action Group Strategies in the Indian Sub-continent", *Development in Practice*, 1(2): 84–96.
———. 1986. *Local Institutional Development*. West Hartford CT: Kumarian Press.
Uphoff, Norman. 1996. "Why NGOs are Not a Third Sector: A Sectoral Analysis with Some Thoughts on Accountability, Sustainability, and Evaluation", in Michael Edwards and David Hulme, (eds.), *Beyond the Magic Bullet: NGO Performance and Accountability in the Post-Cold War World*, pp. 23–39. West Hartford CT: Kumarian Press.
———. 1988. "Assisted Self-Reliance: Working With, Rather than For, the Poor", in Lewis, John P., Guest (ed.), *Strengthening the Poor: What Have We Learned?* pp. 47–59. Washington DC: Overseas Development Council.
Uphoff, Norman, Milton J. Esman and Anirudh Krishna. 1997. *Reasons for Success: Learning from Instructive Experiences in Rural Development*. West Hartford CT: Kumarian Press.
USAID. 1981. *Winning the Cold War: The U.S. Ideological Perspective*. Washington DC: U.S. Government Printing Office.
Uvin, Peter and David Miller. 1996. "Paths to Scaling-up: Strategies for Local Nongovernmental Organizations", *Human Organization*, 55(3): 344–354.
van Blitterswijk, J.D. 1985. *Non Governmental Organizations and Social Forestry in India*. Wageningen, the Netherlands: Department of Forest Management, Wageningen Agricultural University.
Vani, M.S. 1992. *Role of Panchayat Institutions in Irrigation Management: Law and Policy*. New Delhi: Indian Law Institute.
Venkatasamy, D. 1997. Personal communication.
Verma, M.C. and P.G. Menon. 1993. "Role of NGOs: UN and Bilateral Aid", *Proceedings of the International Conference on Sustainable Village-Based Development*, Colorado State University, 27 September–1 October.
Verney, Douglas. 1989. "From Executive to Legislative Federalism? The Transformation of the Political System in Canada and India", *The Review of Politics*, 51(2): 241–263.
Verney, Douglas and Francine R. Frankel. 1986. "India: Has the Trend toward Federalism Implications for the Management of Foreign Policy? A Comparative Perspective", *International Journal*, 41 (June): 572–599.
Vira, Bhaskar, Oliver Dubois, Steven E. Daniels, and George Walker. 1998. "Institutional Pluralism in Forestry: Considerations of Analytical and Operational Tools", *Unasylva*, 49(3), also available from http://www.fao.org/docrep/w8827e/w8827e07.htm, accessed on 2 January 2001.
Vivian, Jessica. 1994. "NGOs and Sustainable Development in Zimbabwe: No Magic Bullets", *Development and Change*, 25(1): 167–193.
Vogel, David. 1986. *National Styles of Regulation: Environmental Policy in Great Britain and the United States*. Ithaca NY: Cornell University Press.
WCED (World Commission on Environment and Development). 1987. *Our Common Future*. New York: Oxford University Press.

Weaver, James H., Michael T. Rock and Kenneth Kusterer. 1996. *Achieving Broad-Based Sustainable Development: Governance, Environment and Growth with Equity.* West Hartford CY: Kumarian Press.

Wellard, Kate and James G. Copestake, (eds.), 1993. *Non-Governmental Organizations and the State in Africa: Rethinking Roles in Sustainable Agricultural Development.* New York and London: Routledge.

Whaites, Alan. 1996. "Let's Get Civil Society Straight: NGOs and Political Theory", *Development in Practice*, 6(3): 240–244.

Wheare, Kenneth C. 1964. *Federal Government*, 4th ed. New York: Oxford University Press.

Whitehead, Sir Alfred North. 1925. *Science and the Modern World.* New York: MacMillan.

———. 1929.*The Function of Reason.* Boston: Beacon Press.

Wood, Donna J. and Barbara Gray. 1991. "Toward a Comprehensive Theory of Collaboration", *The Journal of Applied Sciences*, 27(2): 139–162.

Wood, Geof. 1996. "States without Citizens: The Problem of the Franchise State", in David Hulme and Michael Edwards, (eds.), *NGOs, States and Donors: Too Close for Comfort?* Ch. 5. New York: St. Martin's Press in association with Save the Children.

Yin, Robert K. *Case Study Research: Design and Methods.* Thousand Oaks CA: Sage Publications.

Zadek, Simon and Murdoch Gatward. 1996. "Transforming the Transnational NGOs: Social Auditing or Bust?", in Edwards and Hulme, (eds.), *Beyond the Magic Bullet: NGO Performance and Accountability in the Post-Cold War World.* Hartford CT: Kumarian Press. Ch. 17.

Zubaida, Sami. 1992. "Islam, the State and Democracy: Contrasting Conceptions of Society in Egypt," *Middle East Report*, (November–December): 2–10.

Index

accountability: direction of, 81–82; government, 59, 77, 78, 82; kinds of, 81–82; multiple, 82; mutual, 184–86; of NGOs, 66, 69, 80–82, 90, 91, 105, 112, 113, 184–86, 188
advocacy, 93, 101, 107, 157, 171, 173, 190; skills, 77
agroecology, 96
Air (Preservation and Control of Pollution) Act (1981), 47
Alliband, Terry, 36, 37
Anna University, 126
anti-colonial movements, 68–69, 109
Appleby, Paul, 40
Arora, Balveer, 43n4
Ascher, W., 54–55, 124–25
Auroville Foundation Act, 1988, 130, 131
Auroville Greenwork Resource Centre, 132
Auroville Water Service (AWS), 120, 129–30, 133–37, 155, 156, 157, 158, 159, 164, 165, 169–72, 174, 179
autonomy: of NGOs, 66, 71, 81–82, 91, 94n6, 112, 185, 186; of states, 40, 41; of subsidiary governments, 34

Balwantrai Mehta Commission Report of 1957, 61
Bebbington, Anthony, 75, 87, 91, 96, 97, 98, 99, 113
Bhat, K.V., 161, 176
Bhat, M.K., 21n7

Bhatt, Anil, 65, 94n6
Bose, Nirmal, 41
Bowonder, B., 54
Bratton, Michael, 99–100
Brown, L. David, 92n3
bureaucracy: development, 45–46; as regulatory, 51; rigidity of central, 41

capitalism, 52
Carroll, T., 69
centralization, 40
Chambers, Robert, 146
Charitable and Religious Act, 1920, 111
Chhatre, Ashwini, 45
civil societies, 78, 81n11, 91n2, 190; creation of, 66, 79
Clark, John, 15–16n2, 20, 72–73, 77, 89, 90, 92–93, 107ns8, 9
'clientele politics', 58
common property resources, 51, 126n3, 177, *Also see* natural resources
communication skills, need for, 77
Communist Party of India-Marxist (CPI-M), 63, 114
community development, 36, 37
Concurrent List, 42, 43, 45
Congress Party, 39, 42n3, 109
cooperative movement, 112
corruption, 59, 85, 113, 128, 145
Council for Advancement of People's Action and Rural Technology (CAPART), 46, 111, 112

Council for the Advancement of Rural Technology (CART), 111
critical collaboration, 15, 21, 54, 64, 98, 116, 151, 172; attributes for, 174–81, 184; concept of, 22–23, 100–103, 117, 179, 188; conditions for, 23–24; formal agreements for, 155–56; influencing factors for, 25–26, 27; power-sharing in, 100–102, 104, 151, 154; as a process, 155, 188

decentralization, 33, 35, 37, 45, 61–63, 109n12, 114, 115, 163; of decision-making, 162; of planning, 40, 162
decision-making, 23, 37, 100, 101, 103, 104, 142, 162, 184; rural development, 45; state, 43
demand politics, 58
democracy, 52, 79
democratization, 66, 91n2, 109
Department of Environment (DoE), 48, 49
Department of International Development and Cooperation, Denmark (DANIDA), 146, 164, 165
Department of Non-Conventional Energy Sources (DNES), 167–68
Department of Program Implementation, 46
Department of Rural Development, 46, 137
Devavaram, John, 145–46
development, 15–16n2; and centralization, 57n11; and decentralization, 57n11, 57–59; definition of, 16–18; "demand side" approach to, 77; and environmental problems, 53; national, 19; program/project approach to, 66–67; and protest, 55, 56; public participation in, 55; rural, 15, 18–20, 31, 33, 37, 42, 43, 45, 57, 62; "supply side" approach to local, 77; sustainable, 19 Also see rural development
Dhesi, Antar, 37

District Rural Development Agencies (DRDAs), 46, 126, 127, 128, 137, 163
Dryland Development Board (DLDB), 176

Edwards, M., 67n2, 81, 82, 113
Environment (Protection) Act (1986), 47, 51, 139
environment: and the Constitution, 44; and protest movements, 52, 55–56
environmental appraisal, 107n9
environmental problems, political solutions to, 54–55
environmentalism, history of, 29
Escobar, A., 16n3, 17
European Union, 120

farm forestry, 45
Farrington, J., 75, 83, 87, 89, 90, 91, 96, 97, 98, 99, 113
federalism, 38–42, 45n7, 52, 56, 155
Federation of Voluntary Organisations for Rural Development-Karnataka (FEVORD-K), 160–61
female infanticide, 121
Fernandez, A., 109n11
Fisher, J., 56, 69, 82, 89n1, 97, 105,189
Ford Foundation, 120, 121
Foreign Contributions Regulation Act, 1976, 89, 111, 131
Forest Conservation Act 1980, 45n7, 47
Forty-Second Amendment, 45n7
Fowler, A., 69
Frankel, F., 40
Frantz, T.R., 68
Freire, Paulo, 75

Gablier, Thomas, 170
Gadgil, M., 52–53
Gandhi, Indira, 36, 47
Gandhi, M.K., 61, 62
Gandhi, Rajiv, 62
Garilao, Ernesto, 68
Government of India (GOI): and acknowledgement of NGO sector,

110; and approach to development, 33, 35, 64, 65; and approach to environment, 51–52, 53; and decentralization, 61, 63; Eighth Five-Year Plan of, 37, 110; and expenditure on rural development, 90; as managerial, 53; and popular protest, 56; as regulatory, 51–52, 111, 112, 113
Government of India Act of 1935, 39n2
governments, and authoritarianism, 34
Gram Swaraj, 62
grassroots organizations (GROs), 69, 82, 83, 91n2, 104, 127
grassroots support organizations (GRSOs), 56, 69n5
Gray, Barbara, 28, 100, 101, 102–103, 178, 182
Green Revolution, 35–36
Guha, R., 52–53
Gujarat Energy Development Agency (GEDA), 168
Gupta, Ranjit, 175n4

Hardgrave, R.L., 39n2, 55
Healy, R., 54–55, 124–25
Howarth, David, 16n3
Hulme, D., 67n2, 81, 82, 113
human development, 37

identity, national, 41
illiteracy, 18
income producing microenterprises, 121
income redistribution programs, 36
India: center-state relations in, 39–46; creation of new states in, 43; non-governmentalism in, 68–69, 108; NGOs in, 83–85, *Also see* NGOs; and poverty, 15, 35; and rural population, 15, 35; as a unitary state, 43
Indian Trusts Act of 1982, 111
Industrial Revolution, 17
industrialization, 17, 36
international NGO (INGO), 121, 166

irrigation, 45

Jain, Lakshmi, 94
Jaix, R.B., 58, 72
Jeyakaran, C., 173
Joint Forestry Management programs, 142, 143, 149, 163–64
Judge, Anthony, 67, 70n6

Kadavakurichi Interface Forestry Management Programme, 140–43, 165, 178, 180
Karnataka Panchayat Act of 1983, 62
Khan, Rasheeduddin, 41
Khator, Renu, 47, 48, 49, 50, 51, 52, 57, 168
Khilnani, Sunil, 42n3
Kochanek, S., 39n2, 55
Korten, D., 75–76, 105, 106, 183n1
Kothari, R., 59
Krishna, Sumi, 52, 53n9

Lacey, H., 18n5
Lele, S., 163
'less developed country(ies)' (LDC), 15, 31, 34, 51, 64, 67, 71, 74, 88, 95, 96
Lewis, D.J., 83, 89, 90, 99
Lewis, John P., 175n4
local policy culture (LPC), 25, 57, 157–58

Madras Institute for Development Studies, 136
Maniates, M., 168
Meinzen-Dick, R.S., 119
'membership organizations', 68n4, 69
micro-credit, 121
Ministry of Environment and Forests (MoEF), 49, 139, 150
Ministry of Rural Areas and Employment, 46, 137
minor forest produce, 45
Minor, Robert, 130
Mitra, S.K., 56–57, 58
modernization, 17, 35, 109
Moen, E., 115
Mukhopadhyay, A.K., 41

Najam, Adil, 94, 94n7
National Afforestation and Eco-Development Board (NAEB), 150–51, 155–56, 157, 159, 160, 163, 173, 180
National Committee on Environmental Planning and Co-ordination (NCEPC), 48
National Council of Voluntary Action, 112
National Environmental Council: in Sri Lanka, 107, 107n9
National Program for Improved Chulhas (NPIC), 167–68
national resources: apportionment of, 40
National Wasteland Development Board (NWDB), 140, 162–63
nation-building, 34, 109
natural resources, 18, 42, 51, 53, 54, 118; conservation, 124–29, 141, 142, 177; management, 121, 144, 163; programs, 111–12, *Also see* primary resources
Nehru, Jawaharlal, 35, 36, 39, 51, 109
Nerfin, Marc, 67–68
networking skills: need for, 77
Non-Governmental Organizations (NGO)-Government Organizations (GO) interrelationships, 20, 71, 76, 79, 86, 88–100, 104, 108–16, 117, 118, 129, 130, 135, 136–37, 138–40, 142–43, 145–47, 150–51, 153–81, *Also see* critical collaboration: forms of, 96–100; institutionalization of, 113
NGOs: as action-oriented, 77; and aid donors, 66; as an alternative to government, 63, 64, 71, 96, 110; and autonomy, 66, 71; and charity work, 74; Code of Conduct for, 112, 113, 175, 186, 187; and commitment, 76, 112, 129; concept of, 67–70; and cooperation, 89–91, 92n3, 95–96, 98–99, 101, 107, 108, 123, 131, 135, 140, 159–60, 172, 174, 187, 189; and cooptation, 91, 94, 94n7, 97, 106, 113, 186; and corruption, 113, 186; and disaster relief, 74, 109; and 'first generation' strategies, 75; and focus on empowerment of the poor, 83; and focus on poverty alleviation, 83; and funding, 111–12, 120, 131, 165, 169, 170, 171, 186; history of, 74–79, 110; indigenous, 19–20, 69, 70, 86, 94, 96, 97, 104, 116, 165; informal structure of, 71; as intermediaries between the state and rural society, 78–79; lack of homogeneity in, 119; legitimacy as stakeholders, 104; linkages with GOs, 98, 101; managerial features of, 67n2105–6; nationalist tradition among, 89; and new roles and responsibilities, 76, 78; and policy-making, 123; and popular protest, 56; positive attributes of, 71–72, 79, 166; problems faced by, 127–28; radical, 93; regulation of, 46, 108, 187–88; role of, 37; restriction of, 114; and 'second generation' strategies, 75; as 'service organizations', 69n4, 76; size of, 83, 85, 119, 120, 138, 143, 147–48, 155, 157, 165, 166–72; social action approach of, 84, 85; and strategic competence, 106; and technical competence, 105; and 'third generation' strategies, 75–76; and trust, 78, 86, 107, 189; and welfare activities, 74, 109

operational collaboration, 98, 99, 100, 101

Palni Hills Conservation Council (PHCC), 120, 138–40, 155–57 *passim*, 164–77 *passim*
panchayati raj, 41, 61–63; institutions, 58
Panchayati Raj Act (1993), 62
panchayats, 45, 109n11, 126, 150
parliamentarianism, 52

INDEX 213

Participative Integrated Development of Watersheds (PIDOW), 175–76
Participatory Rural Appraisal (PRA) techniques, 141, 145–47, 156–57, 164, 165, 171, 173, 177
partisan politics, 52
Patel, Sardar Vallabhbhai, 35
Paul, Samuel, 99
People's Action for Development Initiatives (PADI), 111
peoples' associations, 68n4
pollution control, 47
Potter, David, 160
poverty: as an index, 18; reduction programs, 71
President's Rule, 43n4
primary resources, access to, 19
princely states, 38–39
private charities, 29
privatization, 35
problem-solving, 86
Professional Assistance for Development Action (PRADAN), 120–29, 135, 136, 137, 148, 155–87 *passim*

Quizon, A.B., 183n1

Rajasekhar, D., 21n7, 22–23, 89
Reddy, N.L. Narasimha, 21n7, 22–23, 89
Registration of Societies Act, 1860, 108, 111
religious missions, 29
Riddell, R.C., 83
Riker, J.V., 97
Robinson, M., 83, 84, 111
Roy, Bunker, 113
rural development: bottoms-up approach to, 73, 73n7, 183; centralized power in, 59–61, 63; and local elites, 57–59; mainstream approach to, 83–84, 93, 96, 157; and NGOs, 66, 83–84, 85, 111, 116, 121, 122, 158; poverty reduction as key to, 36; programs, 21, 23, 37; as the responsibility of the government, 109; social action approach to, 84, 157; top-down approach to, 73, 73n7, 183, *Also see* development

Sanyal, B., 107
Sarvodaya movement, 109, 109n12
Sastry, K.R., 63
Satish, S., 161, 176
'scaling out', 89, 156
'scaling up', 89n1, 156
self-employment, 121
self-government, 62; village, 61
self-help, 68n4, 74n7, 75, 109, 144; groups, 121; organizations, 104
Sempatti Hill Initiators for Ecological Development (SHIELD), 147–51, 155–56, 157, 159–60, 163, 164, 171, 173, 177, 180
Sen, A., 76
Sen, S.R., 40
Sepoy Mutiny of 1857, 59
Sethi, Harsh, 68, 74–75, 108, 153
Seventy-third Amendment, 1992, 45, 62
Sheth, D.L., 68, 74–75, 108, 153
Singh, H., 62
social action groups, 121
'social auditing', 188
social forestry, 45, 165
social justice, 45, 65
social reform movements, 108
Society for People's Education and Economic Change (SPEECH), 143–47, 155, 156, 157, 159, 164, 165, 171, 173, 177, 180
soil conservation, 45
stakeholders, 21, 85, 100, 101, 102, 103, 154, 155, 174, 179, 184, 186, 188, 189
State List, 40, 42, 43, 45, 46
state: as arbitrator, 56–57, *Also see* Government of India
Swedish International Development Agency (SIDA), 140, 141, 165
Swiss Development Co-operation (SDC), 175–76

Tamil Nadu Association for PRA, 147

Tamil Nadu: availability of water in, 119; deforestation in, 119; Forest Department (TNFD), 140–43, 146, 149–50, 155, 163, 164, 165, 180, 184; maintenance of tanks in, 125–26; Public Works Department (PWD) of, 126, 127, 128; tank irrigation in, 124
Tandon, R., 80, 95, 112
'tank panchayats', 125
Tendler, Judith, 72
Thakur, Ramesh, 44
Tummala, K., 40–41, 58

underdevelopment, 14, 16, 17, 32, 75, 83, 84
underemployment, 18
unemployment, 18
Union List, 42–43

UNESCO, 130
Uphoff, Norman, 68–69n4, 69, 73, 80n10
urbanization, 17, 35

Verney, D., 40
Vivian, Jessica, 71
voluntarism, 65, 94n6, 112, 113
voluntary agencies, 15, 29, 122

Water (Preservation and Control of Pollution) Act (1974), 47
water management, 45
Water Resources Organizations (WROs), 126
Water Users Associations (WUAs), 127, 128
watershed development, 45
Wood, Donna J., 28

About the Author

John M. Riley is a free-lance writer, consultant to a development NGO in India and a small-business manager. His multi-disciplinary academic experience includes teaching geography, political science, and public administration at various universities and community colleges in the USA and overseas. Dr Riley has also been actively involved in research, advocacy, writing, and teaching in the fields of international development and public policy. In addition, he has been associated with various non-governmental organizations and educational groups globally and with several civic bodies in the USA, where he has dealt with rural development and environmental issues. Dr Riley earlier authored the *Community Development Plan, 1988–1992* for the Salt Lake City Corporation.